T0173416

Cambridge Lower Secondary

Maths

STAGE 9: WORKBOOK

Alastair Duncombe, Belle Cottingham, Rob Ellis, Amanda George, Claire Powis, Brian Speed

Series Editor: Alastair Duncombe

Collins

William Collins' dream of knowledge for all began with the publication of his first book in 1819. A self-educated mill worker, he not only enriched millions of lives, but also founded a flourishing publishing house. Today, staying true to this spirit, Collins books are packed with inspiration, innovation and practical expertise. They place you at the centre of a world of possibility and give you exactly what you need to explore it.

Collins. Freedom to teach.

Published by Collins
An imprint of HarperCollins*Publishers*
The News Building
1 London Bridge Street
London
SE1 9GF

HarperCollins*Publishers*
Macken House, 39/40 Mayor Street Upper,
Dublin 1, D01 C9W8, Ireland

Browse the complete Collins catalogue at
www.collins.co.uk

10 9 8 7 6

ISBN 978-0-00-837858-5

This book is produced from independently certified FSC™ paper to ensure responsible forest management.

For more information visit:
www.harpercollins.co.uk/green

British Library Cataloguing in Publication Data
A catalogue record for this publication is available from the British Library.

Authors: Alastair Duncombe, Belle Cottingham, Rob Ellis, Amanda George, Claire Powis, Brian Speed
Series editor: Alastair Duncombe
Publisher: Elaine Higgleton
In-house project editors: Jennifer Hall and Caroline Green
Project manager: Wendy Alderton
Development editors: Anna Cox, Phil Gallagher and Jess White
Copyeditor: Laurice Suess
Proofreader: Tim Jackson
Answer checkers: Tim Jackson and Jouve India Private Limited.
Cover designer: Ken Vail Graphic Design and Gordon MacGlip
Cover illustrator: Ann Paganuzzi
Typesetter: Jouve India Private Limited
Production controller: Lyndsey Rogers
Printed and bound in India by Replika Press Pvt. Ltd.

Acknowledgements

The publishers gratefully acknowledge the permission granted to reproduce the copyright material in this book. Every effort has been made to trace copyright holders and to obtain their permission for the use of copyright material. The publishers will gladly receive any information enabling them to rectify any error or omission at the first opportunity.

p.140 Contains information from Education, Language Spoken and Literacy, accessed on 15/09/20 from https://www.singstat.gov.sg/ which is made available under the terms of the Singapore Open Data Licence version 1.0 {https://data.gov.sg/open-data-licence}; p.140 The World Bank: Mobile cellular subscriptions (per 100 people): International Telecommunication Union (ITU) World Telecommunication/ICT Indicators Database. (CC. by 4.0) https://creativecommons.org/licenses/by/4.0/; p. 146 Contains information from Population and Population Structure, accessed on 15/09/20 from https://www.singstat.gov.sg/ which is made available under the terms of the Singapore Open Data Licence version 1.0 [https://data.gov.sg/open-data-licence]

Cambridge International copyright material in this publication is reproduced under licence and remains the intellectual property of Cambridge Assessment International Education.

Third-party websites and resources referred to in this publication have not been endorsed by Cambridge Assessment International Education.

With thanks to the following teachers and schools for reviewing materials in development: Samitava Mukherjee and Debjani Sen, Calcutta International School; Hawar International School; Adrienne Leisztinger, International School of Budapest; Sujatha Raghavan, Manthan International School; Mahesh Punjabi, Podar International School; Taman Rama Intercultural School; Utpal Sanghvi International School.

Contents

How to use this book

This Workbook accompanies the *Collins Lower Secondary Maths Stage 9 Student's Book* and covers the Cambridge Lower Secondary Mathematics curriculum framework (0862). This Workbook can be used in the classroom or as homework. Answers are provided in the Teacher's Guide.

Every chapter has these helpful features:

- 'Summary of key points': to remind you of the mathematical concepts from the corresponding section in the Student's Book.

- Exercises: to give you further practice at answering questions on each topic covered in the Student's Book. The questions at the end of each exercise will be harder to stretch you.

- 'Thinking and working mathematically' questions (marked as ▼): to help you develop your mathematical thinking. The questions will often be more open-ended in nature.

- 'Think about' questions: encourage you to think deeply and problem solve.

1 Indices, roots and rational numbers

You will practice how to:

- Use positive, negative and zero indices, and the index laws for multiplication and division.
- Understand the difference between rational and irrational numbers.
- Use knowledge of square and cube roots to estimate surds.

1.1 Indices

Summary of key points

Positive, zero and negative indices:

$$2^3 = 2 \times 2 \times 2 = 8$$
$$2^2 = 2 \times 2 = 4$$
$$2^1 = 2$$
$$2^0 = 1$$
$$2^{-1} = \frac{1}{2}$$
$$2^{-2} = \frac{1}{2 \times 2} = \frac{1}{4}$$
$$2^{-3} = \frac{1}{2 \times 2 \times 2} = \frac{1}{8}$$

Index laws for multiplication and division:

$$c^a \times c^b = c^{a+b} \qquad \text{e.g. } 4^5 \times 4^{-9} = 4^{-4}$$
$$c^a \div c^b = c^{a-b} \qquad \text{e.g. } 6^{-2} \div 6^{-7} = 6^5$$

Exercise 1

1 Write the value of each power of 6.

a) $6^0 = $ ………..

b) $6^3 = $ ………..

c) $6^{-2} = $ ………..

d) $6^{-1} = $ ………..

2 Find the missing powers.

a) $3^{\square} = 81$

b) $7^{\square} = \frac{1}{7}$

c) $4^{\square} = 2^{10}$

d) $2^{\square} = 4^{-2}$

3 Write these in order of size, starting with the smallest.

$11^0 \qquad 6^2 \qquad 3^3 \qquad 8^{-1} \qquad 2^{-2}$

……….. ……….. ……….. ……….. ………..

smallest   largest

4 Use each of the numbers −1, −2, −6, 2 and 5 exactly once to complete these statements.

a) $\square^{\square} = 0.2$

b) $8^{\square} = \square^{\square}$

5 Find all the possible pairs of values for the integers a and b where

$$a^b = \frac{1}{16}$$

....

....

6 Decide whether each statement is true or false.

	True	False
$9^{-4} = -9^4$	$\square$	$\square$
$7^{-7} \div 7^{-4} = 7^3$	$\square$	$\square$
$2^{-3} \times 3^7 = 6^4$	$\square$	$\square$
$5^{-7} \times 5^6 = 0.2$	$\square$	$\square$

7 Draw lines to match pairs of equivalent expressions.

$7^6 \times 7^{-5}$	7^8
$7^4 \div 7^4$	7^1
$7^4 \div 7^{-4}$	7^{-1}
$(7^{-4})^2$	7^0
$7^{-6} \times 7^5$	7^{-8}

8 Use positive integers to complete these statements.

a) $\square^{-2} = 4^{-\square}$

b) $\dfrac{1}{\square} = \square^{-3}$

c) $2^5 \times 2^{-\square} = 4^{\square}$

Think about

9 Find other possible answers to question 8 part a.

Describe the relationship between the two missing numbers.

1.2 Rational and irrational numbers

Summary of key points

The **natural numbers** are the positive integers: 1, 2, 3, 4, 5, …

Rational numbers are numbers that can be written as fractions. For example,

$-\dfrac{5}{2}$, $3 \left(= \dfrac{3}{1}\right)$, $0.71 \left(= \dfrac{71}{100}\right)$.

Irrational numbers are numbers that cannot be written as fractions. These include all square roots that are not integers, for example $\sqrt{2}$, $\sqrt{6}$, $\sqrt{21}$.

Exercise 2

1 **a)** Write these labels in the correct places on the Venn diagram.

integers   irrational numbers   natural numbers   rational numbers

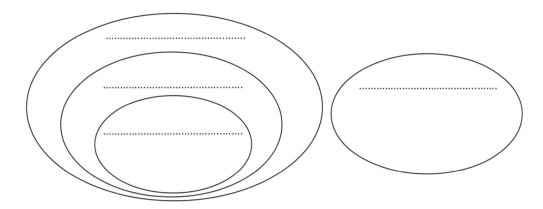

b) Write these numbers in the correct places on the diagram.

17 -5 $5\dfrac{1}{5}$ $\sqrt{25}$ $\sqrt{8}$ 0.0822 $-\dfrac{2}{3}$ π $\sqrt[3]{4}$ $\sqrt[3]{-1}$

2 Angela says she can think of a fraction that is irrational. Is she correct? Explain your answer.

....

....

3 Is each statement true or false?

	True	False
All natural numbers are rational.	☐	☐
The cube root of a cube number is rational.	☐	☐
$\sqrt{8}$ is a rational number.	☐	☐
$0.5\dot{6}\dot{2}$ is a rational number.	☐	☐

4 **a)** Can a recurring decimal be an irrational number? Explain your answer.

....

....

b) Can a terminating decimal be an irrational number? Explain your answer.

....

....

5 **a)** Write a rational number between $\sqrt{7}$ and $\sqrt{10}$. ....

b) Write an irrational number between $\sqrt{7}$ and $\sqrt{10}$. ....

1.3 Estimating surds

Summary of key points

A **surd** is the square root of a number that is not a square number, or the cube root of a number that is not a cube number. Surds are **irrational**.

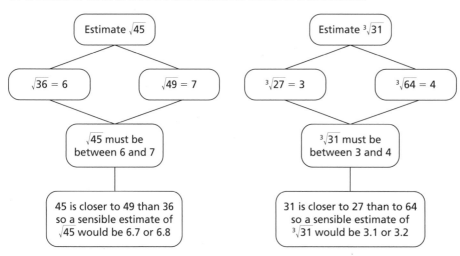

Estimate $\sqrt{45}$

$\sqrt{36} = 6$ $\sqrt{49} = 7$

$\sqrt{45}$ must be between 6 and 7

45 is closer to 49 than 36 so a sensible estimate of $\sqrt{45}$ would be 6.7 or 6.8

Estimate $\sqrt[3]{31}$

$\sqrt[3]{27} = 3$ $\sqrt[3]{64} = 4$

$\sqrt[3]{31}$ must be between 3 and 4

31 is closer to 27 than to 64 so a sensible estimate of $\sqrt[3]{31}$ would be 3.1 or 3.2

Exercise 3

1 Draw a ring around the larger number in each pair.

a) $\sqrt{13}$ or $\sqrt{17}$ 　　 b) 4 or $\sqrt{17}$ 　　 c) $\sqrt[3]{82}$ or $\sqrt[3]{89}$ 　　 d) $\sqrt[3]{102}$ or 5

2 Draw a line to match each root with its best estimate.

a) $\sqrt{80}$ 　　 b) $\sqrt{94}$ 　　 c) $\sqrt[3]{120}$ 　　 d) $\sqrt[3]{60}$

　　4 　　　　　 5 　　　　　 6 　　　　　 9 　　　　　 10

3 Jacques and Philippa each estimate the value of some square roots and cube roots. For each root, write down whose estimate is better.

Question	Jacques' estimate	Philippa's estimate	Name of person with better estimate
$\sqrt{28}$	5	6	
$\sqrt{200}$	15	14	
$\sqrt[3]{11}$	3	2	

4 Draw a ring around the best estimate of each surd.

a) $\sqrt{174}$ 　　　 13 　　　　　 14 　　　　　 15 　　　　　 16

b) $\sqrt{42}$ 　　　 6.0 　　　　　 6.1 　　　　　 6.5 　　　　　 6.9

c) $\sqrt{220}$ 　　　 14.1 　　　　　 14.4 　　　　　 14.6 　　　　　 14.8

5 Estimate to 1 decimal place:

a) $\sqrt{23}$ = ... 　　　　　　 b) $\sqrt{111}$ = ...

c) $\sqrt[3]{12}$ = ... 　　　　　　 d) $\sqrt[3]{121}$ = ...

6 Ginny, Awal and Reggie each try to estimate the cube root of 81.

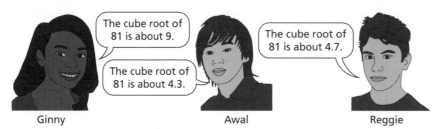

The cube root of 81 is about 9.

The cube root of 81 is about 4.3.

The cube root of 81 is about 4.7.

Ginny　　　　　　　Awal　　　　　　　Reggie

a) Who has the best estimate for $\sqrt[3]{81}$? ....

b) Give a reason for your answer.

....

....

7 a) Write down an estimate of $\sqrt{69}$ to 1 decimal place. ....

b) Use your answer to part **a** to estimate $\sqrt{6900}$ to the nearest whole number. ....

2 Angles

You will practice how to:

- Derive and use the formula for the sum of the interior angles of any polygon.
- Know that the sum of the exterior angles of any polygon is 360°.
- Use properties of angles, parallel and intersecting lines, triangles and quadrilaterals to calculate missing angles.

· ·

2.1 Interior and exterior angles

Summary of key points

The **exterior angles** of any polygon add up to 360°.

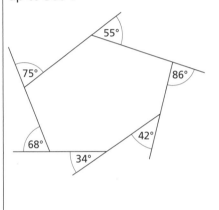

The sum of the **interior angles** of a polygon depends on the number of sides.

If the polygon has *n* sides, the sum of the interior angles is $(n - 2) \times 180°$.

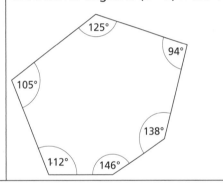

Exercise 1

1 Find the size of the lettered exterior angles.

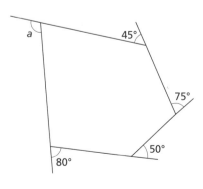

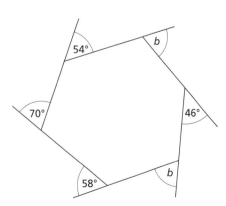

$a = \text{.................}°$

$b = \text{.................}°$

2 Complete the table to show the sum of the interior angles for different polygons.

Number of sides	3	4	5	7	....
Sum of interior angles	180°	.... °	.... °	.... °	1260°

3 Match each diagram to the correct value for *x*. One value of *x* does not match to a diagram.

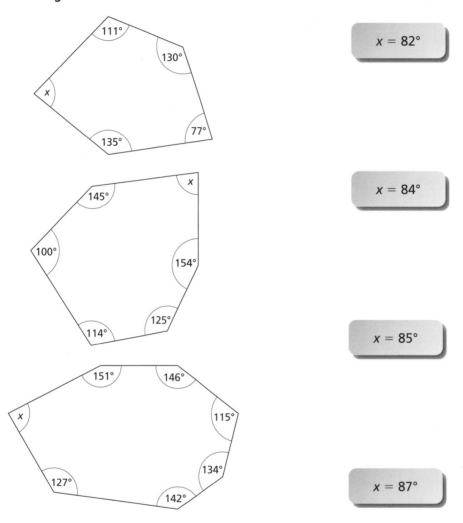

x = 82°

x = 84°

x = 85°

x = 87°

4 Work out the value of *t*.

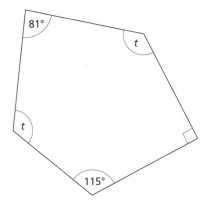

$t =$ .... °

5 **a)** Write down the sum of the exterior angles in an octagon.   .... °

   b) Find the sum of the interior angles of an octagon.   .... °

6 Kamal says that angle *a* is 105°.

Is he correct?   Yes ☐   No ☐

Explain your answer.

....

....

....

....

2.2 Angles of a regular polygon

Summary of key points

In a regular polygon with *n* sides:

- All the exterior angles are equal. They are each $\frac{360°}{n}$.

- All the interior angles are equal. They are each $\frac{(n-2) \times 180°}{n}$.

- The exterior angle + the interior angle = 180°.

A shape can be **tessellated** if it can be used to make a repeating pattern with no overlaps or gaps. A tessellation made from regular polygons is called a **regular tessellation**.

1 Find the size of the lettered angles.

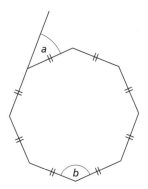

$a = $ .... $^\circ$ $b = $ .... $^\circ$

2 A decagon has 10 sides.

a) Find the size of each exterior angle of a regular decagon.

.... $^\circ$

b) Find the size of each interior angle in a regular decagon.

.... $^\circ$

3 The diagram shows part of a regular polygon with centre *O*.

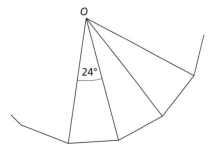

a) Find the size of each interior angle.

.... $^\circ$

b) How many sides are in the complete polygon?

....

4 The exterior angle of a regular polygon is 20°.

Find how many sides the polygon has.

.................°

5 The interior angle of a regular polygon is 150°.

Find how many sides the polygon has.

....

6 The diagram shows a regular pentagon, a regular hexagon and a rhombus.

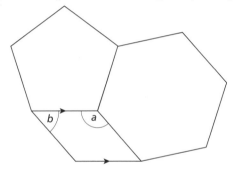

a) Show that $a = 132°$. Give geometrical reasons to support your answer.

....

....

....

....

b) Write down the size of angle b.

$b = $ ....°

7 The exterior angle of a regular polygon is between 7° and 8°.

Find a possible value for the number of sides.

....

2.3 Solving geometrical problems

Summary of key points

Geometrical results that might be needed when giving geometrical reasons include:

Vertically opposite angles are equal.

Alternate angles are equal.

Corresponding angles are equal.

Angles on a straight line add up to 180°.

Angles around a point add up to 360°.

Angles in a triangle add up to 180°.

Base angles in an isosceles triangle are equal.

Angles in a quadrilateral add up to 360°.

Opposite angles in a parallelogram are equal.

A kite has a line of symmetry and so has one pair of equal angles.

Exercise 3

1 **In the diagram, *DE* = *DC*.**

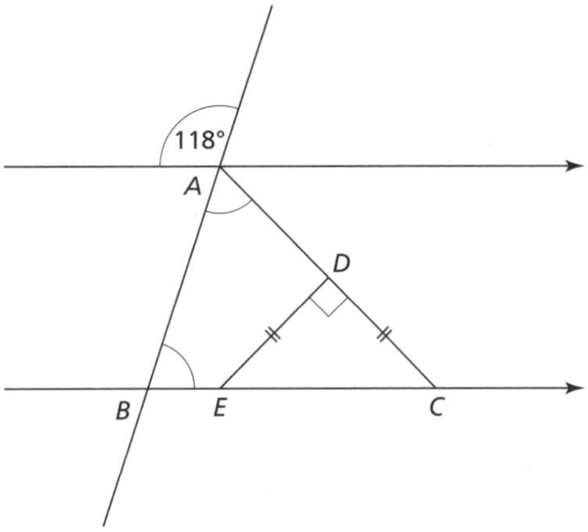

a) Find the size of angle *ABE*.

Give a reason for your answer.

ABE = ……………. ° because ……………………………….

……………………………….

……………………………….

b) Calculate the size of angle *DAB*.

DAB = ………………. °

2 A quadrilateral is drawn inside a parallelogram *ABCD*.

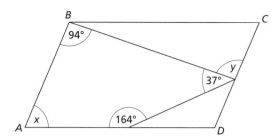

a) Work out the size of angle *x*. Give a reason for your answer.

$x =$ ....°

....

b) Calculate the size of angle *y*.

$y =$ .... °

3 Find the size of the angles marked *x* and *y*.

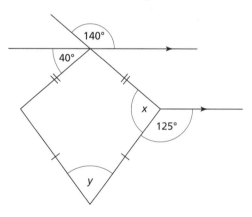

$x =$ ....° $y =$ ....°

4 *ABC* is a right angled triangle.

Fiona says that triangle *ADE* is isosceles.

Is Fiona correct?   Yes [ ]   No [ ]

Show how you worked out your answer.

....

....

....

....

3 Collecting and organising data

You will practice how to:

- Select, trial and justify data collection and sampling methods to investigate predictions for a set of related statistical questions, considering what data to collect, and the appropriateness of each type (qualitative or quantitative; categorical, discrete or continuous).
- Explain potential issues and sources of bias with data collection and sampling methods, identifying further questions to ask.
- Record, organise and represent categorical, discrete and continuous data. Choose and explain which representation to use in a given situation:
  - o Venn and Carroll diagrams
  - o tally charts, frequency tables and two-way tables.

· ·

3.1 Data collection

Summary of key points

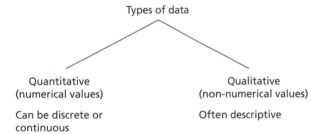

Types of data

Quantitative (numerical values)

Can be discrete or continuous

Qualitative (non-numerical values)

Often descriptive

Some data collection methods can give information that is **biased**. **Bias** can result if some groups of people are over-represented or under-represented in a survey.

Exercise 1

 1 **Draw a ring around the correct phrase to describe each of these variables.**

    a) Number of hours of sunshine   Quantitative data   Qualitative data

    b) Colour of flower   Quantitative data   Qualitative data

    c) Radius of wheel (in cm)   Quantitative data   Qualitative data

    d) Country of birth   Quantitative data   Qualitative data

    e) Type of painting (landscape, portrait, …)   Quantitative data   Qualitative data

2 Juan is a doctor. He wants to collect information about the mass of his patients each time they come for an appointment.

He decides that he could collect this information using one of these methods.

Method A	**Method B**
Ask each patient for their mass in kilograms.	Ask each patient to step onto some scales during the appointment and record the mass.

a) Write down the type of data (qualitative or quantitative) that Juan would be collecting using Method A.

....

b) Give a reason why Juan may want to use Method B.

....

....

c) Give a reason why Juan may **not** want to use Method B.

....

....

3 Bipul grows strawberries. He wants to record the amount of strawberries he picks each day.

Here are two variables he considers recording.

Variable A	**Variable B**
Mass of strawberries picked (in kg)	Number of strawberries picked

a) What type of data is Variable A?   ....

b) Which of the variables would you recommend Bipul to use? Give a reason for your answer.

....

....

4 Fran wants to know how many books men and women read in a month.

a) Write down the two variables that Fran will need to record for her investigation. For each one, tick to show whether the data are qualitative or quantitative.

(1) .... Qualitative ☐   Quantitative ☐

(2) .... Qualitative ☐   Quantitative ☐

b) She collects data by asking some people in a bookshop. Give a reason why this might not give her reliable results.

....

....

5 Jane wants to compare the masses of new-born baby boys with the masses of new-born baby girls.

a) Write down the two sets of data Jane will need for her investigation.

(1) .... (2) ....

b) She considers asking 10 of her friends to tell her their mass when they were born. Why might this not give her reliable results?

....

....

c) Suggest a better source for the data.

....

6 3000 people live in a village. Monty wants to find out how people in the village feel about a new skatepark that is planned for the village.

He writes a questionnaire and gives it to 50 people who live nearest to the site of the skatepark.

Give a reason why the data he collects may be biased.

....

....

7 Isuri wants to find out the views of people in her town about the cost of cinema tickets.

She designs a questionnaire and gives it to people as they leave the cinema after a film. Her questionnaire contains the following question:

Don't you agree that cinema tickets are affordable?

Write down two possible sources of bias in Isuri's investigation.

(1) ....

....

(2) ....

....

8 Paul wants to get a sample of members of a golf club. He has a list of the members of the club in the year 2000.

He uses this list to select a sample of size 40.

Give a reason why his sample may not be representative of members of the golf club.

....

....

Think about

9 Aisla wants to know whether 14-year-old students in her school spend more time playing on a computer than 11-year-old students in her school.

Discuss how Aisla could collect her data. You should include:
- the sets of data she will need
- how she could collect these data
- the sample sizes required
- the potential sources of bias.

3.2 Frequency tables and Venn diagrams

Summary of key points

Data from a survey or experiment can be recorded using a data collection sheet.

For example, this data collection sheet could be used to record the heights of some boys and girls.

Height (*h* cm)	$130 \leq h < 140$	$140 \leq h < 150$	$150 \leq h < 160$	$160 \leq h < 170$	$170 \leq h < 180$
Boys					
Girls					

Data collection sheets and frequency tables sometimes need to be redesigned (for example by changing the width of intervals) to improve the usefulness of the data.

1 Sue wants to find out the distance that workers in her office travel to work. She knows that no one travels more than 50 km.

Complete the row headings of this data collection sheet so that it meets Sue's needs. All intervals should have equal width.

Distance, x (km)	Tally	Total
.... $< x \leq$		
.... $< x \leq$		
.... $< x \leq$		
.... $< x \leq$		
.... $< x \leq$		

2 Shazia designs a data collection sheet to record the number of passengers on each bus arriving at a bus station.

She tests her sheet by recording the number of passengers on a small number of buses.

Number of passengers	Tally	Total
Under 15	‖‖‖ ‖‖	
15–19	‖‖‖	
20–24	‖	
25–29		
30–34		

Design a new data collection sheet that will better meet Shazia's needs.

Number of passengers	Tally	Total

3 Henry is running two activity weeks for children during the school holidays. He asks a small number of the children taking part in each week which activity they would most like to do.

He records the responses on a data collection sheet.

Activity	Week 1	Week 2
Painting		
Model making	I	
Football	III	⦀⦀
Tennis		I
Other	⦀⦀ III	⦀⦀ I
Total	12	

What would you recommend Henry should do to his data collection sheet before he collects the information from all the children attending?

....

....

4 Nikki records the number of passengers getting off 20 trains that arrive at a station.

3	7	11	13	17	12	2	8	0	6
14	12	10	9	13	14	7	3	5	4

a) Nikki uses this frequency table to summarise her data. Complete Nikki's table.

Number of passengers	Tally	Total
0–9		
10–19		
20–29		
30–39		

b) Why are Nikki's class intervals not appropriate?

....

c) Design and complete a better frequency table for Nikki's data. Your class intervals should have equal widths. (You do not need to use all the rows in the table.)

Number of passengers	Tally	Total

5 The amounts of money spent by 20 customers in a supermarket are:

$15.45	$29.34	$41.25	$56.39	$39.17
$4.18	$36.50	$78.45	$63.25	$70.45
$45.30	$55.67	$67.64	$54.19	$163.44
$35.33	$43.28	$45.29	$29.60	$34.58

a) Complete the frequency table to show the information.

Money spent ($)	Tally	Total
0–39.99		
40–79.99		
80–119.99		
120–159.99		
160–199.99		

b) Complete this alternative frequency table to show the information.

Money spent ($)	Tally	Total
0–19.99		
20–39.99		
40–59.99		
60–79.99		
80 or more		

c) Why might the second table be more suitable for these data?

....

....

6 Sam sorts the books on his bookcases. He records the type of cover, type of book and number of pages.

The Venn diagram shows some of the information.

```
fiction                    hard cover

        7
  13          5
        9
           10
     15
 6         11

    less than 200 pages
```

a) Find the number of hard cover books that have less than 200 pages.

....

b) Find the number of fiction books that do not have a hard cover.

....

c) Find the fraction of fiction books that have less than 200 pages.

....

d) Explain what the number 6 represents in the Venn diagram.

....

....

7 Stacey is a baker. The table shows information about the cakes that Stacey makes one month.

	Chocolate flavour		Not chocolate flavour	
	Decorated	Not decorated	Decorated	Not decorated
Round	15	7	18	8
Not round	9	5	11	14

a) Put the information in the table into the Venn diagram.

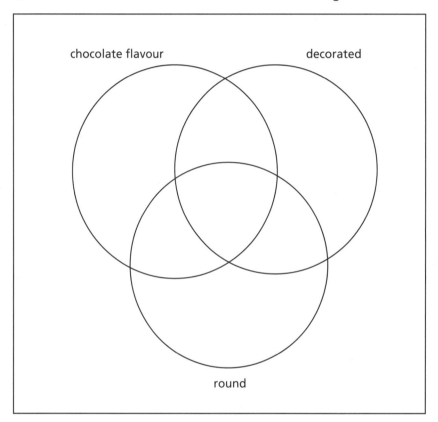

b) Find the number of cakes Stacey makes that are chocolate flavour.

....

c) Show that more than 60% of the decorated cakes are round.

4 Standard form

You will practice how to:

- Multiply and divide integers and decimals by 10 to the power of any positive or negative number.
- Understand the standard form for representing large and small numbers.

4.1 Multiplying and dividing by powers of 10

Summary of key points

The table shows some powers of 10.

10^{-3}	10^{-2}	10^{-1}	10^0	10^1	10^2	10^3
$\frac{1}{1000} = 0.001$	$\frac{1}{100} = 0.01$	$\frac{1}{10} = 0.1$	1	10	100	1000

Multiplying by powers of 10	Dividing by powers of 10
Examples:	Examples:
$0.054 \times 10^4 = 0.054 \times 10\,000$	$12.5 \div 10^3 = 12.5 \div 1000$
$= 540$	$= 0.0125$
$63 \times 10^{-2} = 63 \times \frac{1}{100}$	$0.07 \div 10^{-1} = 0.07 \div \frac{1}{10}$
$= 63 \div 100$	$= 0.07 \times 10$
$= 0.63$	$= 0.7$

Exercise 1

1 Are these statements true or false?

	True	False
$10^{-1} = 0.1$	☐	☐
$10^5 > 5 \times 10$	☐	☐
$2 \times 10^{-1} < 0$	☐	☐

2 Calculate:

a) $12 \times 10^3 =$ ....

b) $0.302 \times 10^2 =$ ....

c) $0.0072 \times 10^4 =$ ....

d) $216 \div 10^2 =$ ....

e) $10.4 \div 10^3 =$ ....

f) $0.46 \div 10^1 =$ ....

3 Tick (✓) the correct statements.

| $21 \times 10^{-1} = -210$ | $7 \times 10^{-1} = 0.7$ | $760 \times 10^{-2} = 76\,000$ |

| $0.086 \div 10^{-1} = 0.0086$ | $0.26 \div 10^{-3} = 26$ | $32 \div 10^{-2} = 3200$ |

4 Complete these calculations.

a) $34 \div 10^{-1} =$ ....

b) $85 \times 10^{-2} =$ ....

c) $0.049 \times 10^{-1} =$ ....

d) $0.27 \div 10^{-3} =$ ....

e) .... $\times 10^4 = 2560$

f) .... $\div 10^3 = 0.0775$

5 Use the numbers on the cards below exactly once to complete the four number statements.

| 10^{-2} | 10^{-1} | 10^2 | 10^3 |

$4.2 \times \square = 4200$ $1.75 \div \square = 175$ $3.5 \div \square = 0.035$ $0.098 \times \square = 0.0098$

6 Complete these statements.

a) $6500 \times 10^{-2} =$ .... $\times 10^2$

b) $7900 \div 10^3 =$ .... $\times 10$

7 A plane travels 10^3 km in 1 hour. The distance around the Earth is $40\,000$ km. Lottie says it would take the plane 4 hours to travel around the Earth. Is Lottie correct? Explain your answer.

....

....

Think about

8 Write 10 different calculations with the answer 60. Each calculation should involve multiplying or dividing a number by a power of 10.

4.2 Standard form

Summary of key points

Any number can be written in standard form:

$$a \times 10^b$$

where

a is greater than or equal to 1 and less than 10

b is an integer.

Examples:

$723\,000 = 7.23 \times 10^5$

$0.00021 = 2.1 \times 10^{-4}$

Exercise 2

1 Draw a ring around the numbers that are written in standard form.

7.00×10^{12} $\qquad$ 15×10^{-6} $\qquad$ 6.1×10^1

$2.81 \div 10^{-7}$ $\qquad$ 5×100^3 $\qquad$ 4.57×10^{-33}

2 Complete the gaps in the table.

930 000	=	$9.3 \times 100\,000$	=	9.3×10^5
	=	$7 \div 100\,000$	=	7×10^{-5}
4 520 000	=		=	
	=	$5.55 \times 10\,000$	=	
	=		=	3×10^{-6}

3 Write each number in standard form.

a) $6\,300\,000 =$ ....

b) $0.07 =$ ....

c) $50\,005 =$ ....

d) $0.00000621 =$ ....

e) fifty-two thousand = ...

f) two million = ...

4 Write each number as an ordinary number.

a) $6.1 \times 10^3 =$ ....

b) $8 \times 10^{-5} =$ ....

c) $9.8 \times 10^1 =$ ....

d) $3.33 \times 10^0 =$ ....

e) $7.034 \times 10^6 =$ ....

f) $1.1 \times 10^{-4} =$ ....

5 Antonio says, '8×10^5 is greater than 4×10^6 because 8 is greater than 4.'
Is he correct? Explain your answer.

....

....

6 Anjali and Owen want to write 30×10^6 in standard form.
Anjali say it is 3×10^7. Owen says it is 3×10^5.
Explain which answer is correct.

....

....

5 Expressions

You will practice how to:

- Understand that the laws of arithmetic and order of operations apply to algebraic terms and expressions (four operations and integer powers).
- Understand how to manipulate algebraic expressions including:
  - expanding the product of two algebraic expressions
  - applying the laws of indices
  - simplifying algebraic expressions.
- Understand that a situation can be represented either in words or as an algebraic expression, and move between the two representations (including squares, cubes and roots).

5.1 Substitution

Summary of key points

Remember to use the correct order of operations when your substitute into algebraic expressions.

For example:

$A = \frac{1}{2}h(a + b)$

Find the value of A if $h = 5$, $a = 9.2$ and $b = 7.4$.

$A = \frac{1}{2}h(a + b)$

$\quad = \frac{1}{2} \times 5 \times (9.2 + 7.4)$

$\quad = \frac{1}{2} \times 5 \times 16.6$ (Remember, brackets first.)

$\quad = \frac{1}{2} \times 83$

$\quad = 41.5$

Find the value of $(k + 15h)^2$ when $h = -0.4$ and $k = 20$.

$(k + 15h)^2 = (20 + 15 \times -0.4)^2$

$\quad\quad\quad\quad = (20 - 6)^2$

$\quad\quad\quad\quad = (14)^2$

$\quad\quad\quad\quad = 196$

Exercise 1

1 If $a = 5$, $b = 12$ and $c = -3$, find the value of:

a) $2c^2$

b) $bc - 3a$

....

c) $a^2 + 8c$

....

d) $(b - 2a)(b + 2a)$

....

....

2 If $e = -4.5$, $f = 8$ and $g = 6$, find the value of:

a) $fg - 2e$

b) $\dfrac{5g - 6}{8}$

....

....

c) $e^2 + f^2 - g^2$

d) $e(f + g)$

....

....

3 If $p = 1.2$, $q = 0.9$ and $r = 0.7$, find the value of:

a) $4(p + 2q)$

b) $\left(\dfrac{1}{4}p - r\right)^2$

....

....

c) $100pq - 12$

d) $\sqrt{3pq} + r$

....

....

4 A formula used in science is $v = u + at$.

Find:

a) v when $u = -20$, $a = 10$ and $t = 2.8$

$v = $ ....

b) t when $v = 19$, $u = 10$ and $a = 2$

5 A formula in science is $E = \frac{1}{2}mv^2$.

Match the values of m and v with the corresponding value of E.

A	B	C	D
$m = 8$ $v = 5$	$m = 16$ $v = 0.5$	$m = 0.5$ $v = 16$	$m = 2.5$ $v = 8$

W	X	Y	Z
$E = 2$	$E = 64$	$E = 80$	$E = 100$

A = ... B = ... C = ... D = ...

6 $s = \dfrac{(v - u)(v + u)}{2a}$ is a scientific formula.

Find:

a) s when $v = 10\frac{3}{4}$, $u = 1\frac{1}{4}$ and $a = 3$

$s = ...$

b) s when $v = 8.5$, $u = -3.5$ and $a = 1.25$

$s = ...$

7 Here is a formula from science:

$\dfrac{1}{u} + \dfrac{1}{v} = \dfrac{1}{f}$

Find f when $u = 4$ and $v = 12$.

$f = ...$

8 Find the value of $24x^{-3} + \frac{1}{4}y^2$ when

a) $x = 2$ and $y = 6$　　　　　　　b) $x = 0.1$ and $y = -2$

………..　　　　　　　　………..

Think about

9 Danni is substituting pairs of values into the expression $\sqrt{\dfrac{2x+6}{y-2}}$.
He finds that some pairs of values of x and y will not give him an answer.
Why is this?

5.2 Expanding a pair of brackets

Summary of key points

When you **expand brackets**, you multiply out the terms in brackets, then simplify your answer.

If you multiply two linear brackets together, your answer is a **quadratic expression**.

Method 1

Expand $(x - 5)(x + 3)$

Multiply each term in the first bracket by each term in the second bracket.

$(x - 5)(x + 3)$

$(x - 5)(x + 3) = x^2 + 3x - 5x - 15$

collect x terms together

$= x^2 - 2x - 15$

Method 2

Expand $(x + 4)(x + 2)$

	x	4
x	x^2	$4x$
2	$2x$	8

$(x + 4)(x + 2) = x^2 + 4x + 2x + 8$

$= x^2 + 6x + 8$

1 **Expand and simplify:**

   a) $(x + 3)(x + 5)$ **b)** $(x + 8)(x + 2)$

               ....

   c) $(x + 4)(x + 8)$ **d)** $(x + 9)(x + 4)$

               ....

2 **Match the equivalent expressions.**

$(x + 2)(x + 6)$	$x^2 + 4x + 3$
$(x + 1)(x + 3)$	$x^2 + 7x + 12$
$(x + 7)(x + 3)$	$x^2 + 8x + 12$
$(x + 5)(x + 4)$	$x^2 + 9x + 20$
$(x + 4)(x + 3)$	$x^2 + 10x + 21$

3 **Expand and simplify:**

   a) $(x - 5)(x + 4)$ **b)** $(x - 7)(x + 4)$

               ....

c) $(n + 4)(n - 6)$

d) $(x + 2)(x - 2)$

....

....

e) $(y - 4)(y - 10)$

f) $(c - 5)(c + 5)$

....

....

4 Complete these quadratic expansions.

a) $(x + 4)^2 = x^2 + \ldots\ldots\ldots x + \ldots\ldots\ldots$

b) $(x - 7)^2 = x^2 - \ldots\ldots\ldots x + \ldots\ldots\ldots$

c) $(x - 9)^2 = x^2 - \ldots\ldots\ldots x + \ldots\ldots\ldots$

5 Kira is trying to expand $(x + 6)(x - 2)$.

Her working is shown below.

$(x + 6)(x - 2) = x^2 + 2x - 6x - 12 = x^2 - 4x - 12$

Has Kira expanded the brackets correctly? Explain your answer.

....

....

6 Complete these statements.

a) $(x + 1)(x + \ldots\ldots\ldots) = x^2 + 9x + 8$

b) $(x + 3)(x + \ldots\ldots\ldots) = x^2 + \ldots\ldots\ldots x + 30$

c) $(x - 4)(x + \ldots\ldots\ldots) = x^2 + \ldots\ldots\ldots x - 36$

d) $(x - 7)(x - \ldots\ldots\ldots) = x^2 - \ldots\ldots\ldots x + 35$

Think about

7 Find two linear expressions that multiply together to give $x^2 + 12x + 35$.

Summary of key points

a^5 means $a \times a \times a \times a \times a$.

The number 5 is known as a **power** or **index**. It tells you how many times a is multiplied by itself.

When you **multiply**, add the powers.	When you **divide**, subtract the powers.
$a^n \times a^m = a^{n+m}$	$a^n \div a^m = a^{n-m}$
Example: $m^2 \times m^4 = m^6$	Example: $n^{12} \div n^4 = n^8$

Exercise 3

1 Write these expressions in power form.

a) $t \times t \times t \times t \times t \times t$ ....

b) $k \times k \times k \times k \times k \times k \times k$ ....

2 Match the equivalent expressions.

$n^4 \times n^3$	$n^5 \times n$	$n^2 \times n^3$	$n^6 \times n^2 \times n^4$

n^5	n^6	n^7	n^{12}

3 Write each expression as a single power of x.

a) $x^5 \times x^6 =$ ....

b) $x^7 \times x \times x^3 =$ ....

c) $x^{12} \div x^6 =$ ....

d) $\dfrac{x^3 \times x^4}{x^5} =$ ....

4 Use ticks (✓) and crosses (✗) to mark Priti's homework.

1. $p^4 \times p^5 = p^9$ 4. $s^{12} \div s^6 = s^2$

2. $q^3 \times q = q^3$ 5. $u^7 \div u^3 = u^4$

3. $r^3 \times r^3 \times r^2 = r^{18}$ 6. $v^9 \div v^3 = v^6$

5 Complete these statements.

a) $t^{\square} \times t^4 = t^8$ b) $t^3 \times t^{\square} = t^{11}$

c) $\dfrac{t^{14}}{t^{\square}} = t^5$ d) $\dfrac{t^{10}}{t^{\square} \times t} = t^2$

6 Explain why $p^4 \times q^3$ cannot be simplified.

....

....

7 Toni thinks that $(d^4)^3$ simplifies to d^7.

Treya thinks that it simplifies to d^{12}.

Who is correct?   ....

Explain how you know.

....

....

8 Draw a ring around the pair(s) of numbers that give the same value for x^y and y^x.

| $x = 1$ | $x = 2$ | $x = 4$ | $x = 0$ | $x = 3$ |
| $y = 4$ | $y = 3$ | $y = 2$ | $y = 5$ | $y = -3$ |

Think about

9 Make up 10 questions involving multiplication and division of indices that simplify to y^{10}.

Summary of key points

To manipulate **algebraic fractions**, you apply the same rules as you use for arithmetic fractions.

Examples:

Work out: $\dfrac{x}{4} + \dfrac{3y}{10}$

First find a suitable common denominator, in this case 20.

$\dfrac{x}{4} = \dfrac{5x}{20}$ and $\dfrac{3y}{10} = \dfrac{6y}{20}$

$\times 5$ $\times 2$

So, $\dfrac{x}{4} + \dfrac{3y}{10} = \dfrac{5x}{20} + \dfrac{6y}{20} = \dfrac{5x + 6y}{20}$

Simplify $\dfrac{12x + 4}{4}$

$\dfrac{12x + 4}{4} = \dfrac{4(3x + 1)}{4}$ factorising the numerator

$= \dfrac{4(3x + 1)}{4}$ cancelling the common factors

$\left(\dfrac{4}{4} = 1\right)$

$= 3x + 1$

Exercise 4

1 **Work out:**

a) $\dfrac{x}{7} + \dfrac{4x}{7}$ ....

b) $\dfrac{9y}{5} - \dfrac{3y}{5}$ ....

c) $\dfrac{4}{r} + \dfrac{3}{r}$ ....

2 **Is each statement true or false?**

	True	False
$\dfrac{n}{7} + \dfrac{n+1}{7} = \dfrac{2n+1}{7}$	☐	☐
$\dfrac{a}{8} - \dfrac{b}{2} = \dfrac{a-b}{6}$	☐	☐
$\dfrac{7n}{10} - \dfrac{n}{5} = \dfrac{n}{2}$	☐	☐
$\dfrac{n}{3} + \dfrac{m}{4} = \dfrac{3n + 4m}{12}$	☐	☐

3 **Calculate:**

a) $\dfrac{t}{2} - \dfrac{t}{8}$

b) $\dfrac{n}{15} + \dfrac{m}{3}$

c) $\dfrac{n}{2} + \dfrac{3}{10}$

....

4 Complete these statements.

a) $\dfrac{\Box}{9} + \dfrac{1}{3} = \dfrac{2n+3}{9}$

b) $\dfrac{\Box}{n} - \dfrac{2h}{n} = \dfrac{3h}{n}$

c) $\dfrac{t}{4} + \dfrac{u}{\Box} = \dfrac{3t+\Box}{12}$

5 Complete these calculations.

a) $\dfrac{12a - 9}{3} = \dfrac{3(\ldots - 3)}{3} = \ldots$

b) $\dfrac{10b + 25c}{5} = \dfrac{\ldots(2b + \ldots)}{5} = \ldots$

c) $\dfrac{3x^2 + 2x}{x} = \dfrac{\ldots(\ldots + 2)}{x} = \ldots$

6 Simplify these fractions.

a) $\dfrac{7x + 21}{7}$

b) $\dfrac{24x - 16}{8}$

c) $\dfrac{12x + 15y}{3}$

………..   ………..   ………..

d) $\dfrac{60a - 24}{12}$

e) $\dfrac{4m^2 + 5m}{m}$

f) $\dfrac{4x^2 + 6x}{2x}$

………..   ………..   ………..

7 Simplify these fractions.

a) $\dfrac{4x}{5} \times \dfrac{2}{3}$

b) $\dfrac{x}{2} \times \dfrac{x^2}{3}$

c) $\dfrac{3m}{4} \times \dfrac{2m}{9}$

………..   ………..   ………..

8 Draw a ring around the calculations that simplify to $2x + 3$.

$\dfrac{2}{x} \times \dfrac{2}{3}$ $\dfrac{150x + 225}{75}$ $\dfrac{2x^2 + 3x}{x}$ $\dfrac{2x}{5} + \dfrac{3}{5}$ $\dfrac{6x + 9}{3}$ $\dfrac{4x^2 + 12}{3}$

Think about

9 Write as many algebraic fraction calculations as you can that give an answer of $\frac{3n}{4}$.

5.5 Constructing expressions

Summary of key points

Algebraic expressions are often used to represent real-life situations.

For example:

The diagram shows a rectangle *ABCD* joined to a square *EFGH*.

Find an expression for the area of the entire shape.

Area of rectangle $ABCD = x(x + 8)$

$$= x^2 + 8x$$

Area of square $EFGH = x^2$

Total area of the shape $= x^2 + 8x + x^2$

$$= 2x^2 + 8x \text{ cm}^2$$

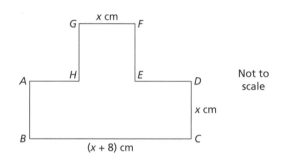

Not to scale

Exercise 5

1 Amir puts plants into four containers.

a) Complete the table by writing expressions for the number of plants in each container.

Container	Description of number of plants	Expression for number of plants (in terms of p)
Red	This container has p plants.	p
Blue	The blue container has 2 more plants than the red container.	
Green	The green container has twice as many plants as the red container.	
Yellow	The yellow container has 1 less plant than the green container.	

b) Find and simplify an expression for the total number of plants Amir puts into the containers.

....

2 Match each 'think of a number' problem to the correct expression.

I think of a number, n. I multiply it by 5. I subtract 2. I square the result.	I think of a number, n. I square the number. I subtract 2. I multiply by 5.	I think of a number, n. I subtract 5. I square the expression. I multiply by 2.
$2(n - 5)^2$	$(5n - 2)^2$	$5(n^2 - 2)$

3 The diagram shows a shape *ABCDEF* formed from two rectangles.

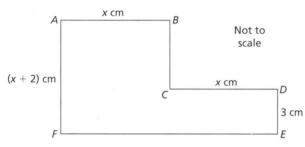

Not to scale

Write and fully simplify an expression for the area of the shape.

.... cm²

4 Write expressions for these 'think of a number' puzzles for any number, n.

a) I think of a number, n.
I square the number.
I multiply by 5.
I then subtract 6.

b) I think of a number, n.
I add 8 to the number.
I divide the answer by 2.
I then square root the expression.

....

5 A rectangle has lengths as shown.

$(x - 4)$ cm

$(x + 10)$ cm

a) Write an expression for the area of the rectangle. Write your expression without brackets.

.... cm²

b) If $x = 7$, what are the dimensions of the rectangle?

Length = ... cm

Width = ... cm

c) If $x = 7$, what is the area of the rectangle? Show your working:

   i) from the dimensions   **ii)** from your algebraic expression.

....

6 Complete this 'think of a number 'puzzle:

I think of a number, n.

I ....

I then ....

Finally, I ....….....

My finishing number is $3(n - 2)^2$.

7 To find the volume of a square-based pyramid, you find the area of the base, then multiply it by the height and multiply your answer by $\frac{1}{3}$.

Use this information to write an expression for the volume of this pyramid.

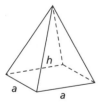

....

6

Transformations

You will practice how to:

- Transform points and 2D shapes by combinations of reflections, translations and rotations.
- Identify and describe a transformation (reflections, translations, rotations and combinations of these) given an object and its image.
- Recognise and explain that after any combination of reflections, translations and rotations the image is congruent to the object.
- Enlarge 2D shapes, from a centre of enlargement (outside, on or inside the shape) with a positive integer scale factor. Identify an enlargement, centre of enlargement and scale factor.
- Analyse and describe changes in perimeter and area of squares and rectangles when side lengths are enlarged by a positive integer scale factor.

6.1 Describing transformations

Summary of key points

It is important to give a full description of a transformation.

Translation – give the vector $\begin{pmatrix} a \\ b \end{pmatrix}$ where a is the number of units moved horizontally and b is the number of units moved vertically.	**Reflection** – give the equation of the mirror line.	**Rotation** – give the angle and direction of rotation as well as the coordinates of the centre of rotation.

Exercise 1

1 Complete the description of each transformation that maps shape *P* to *Q*.

a)

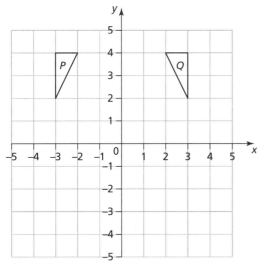

Reflection in ....

b)

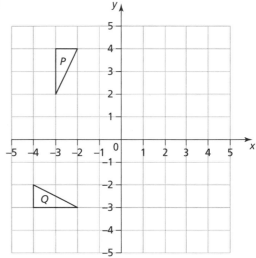

Rotation by .... ° anticlockwise,

centre (.........., ...)

c)

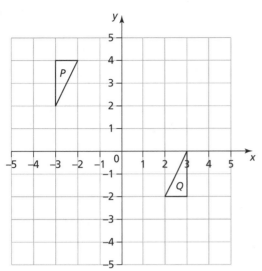

Rotation of ....

centre (........... , ...)

d)

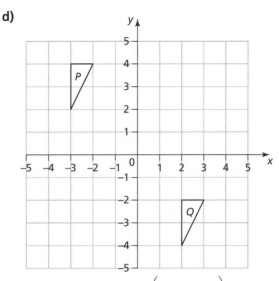

Translation by vector $\begin{pmatrix} .... \\ .... \end{pmatrix}$

2 **Fully describe each transformation.**

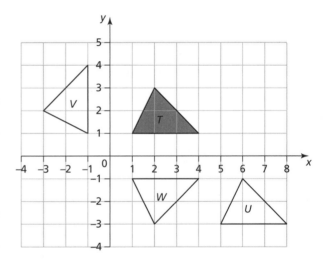

a) The transformation that maps triangle
   T to triangle *U* is

   ....

   ....

b) The transformation that maps triangle
   T to triangle *V* is

   ....

   ....

c) The transformation that maps triangle *T* to triangle *W* is

   ....

3 **Describe fully the single transformation that maps:**

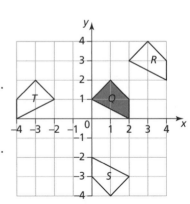

a) quadrilateral *Q* to quadrilateral *R*

   ....

b) quadrilateral *Q* to quadrilateral *S*

   ....

c) quadrilateral *Q* to quadrilateral *T*

   ....

4 The diagram shows flags *F*, *G* and *H*.

a) Give a reason why the transformation from *F* to *G* cannot be a translation.

....

....

b) Describe the single transformation that maps *F* to *H*.

....

....

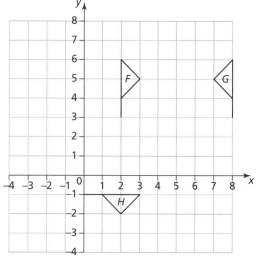

Summary of key points

In the diagram, the trapezium is:

- first translated with vector $\begin{pmatrix} 6 \\ 0 \end{pmatrix}$
- then rotated by 90° clockwise, centre (0, 0).

The single transformation that maps the object (starting shape) to the image (final shape) is a rotation by 90° clockwise, centre (−3, −3).

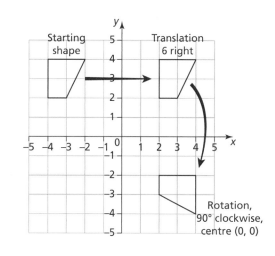

Exercise 2

1 The diagram shows a shape *P* on a grid.

a) Translate shape *P* with vector $\begin{pmatrix} 4 \\ -2 \end{pmatrix}$. Label the image *Q*.

b) Translate shape *Q* with vector $\begin{pmatrix} -5 \\ -5 \end{pmatrix}$. Label the image *R*.

c) Draw a ring around the single transformation that maps *P* directly onto *R*.

translation   reflection

rotation   enlargement

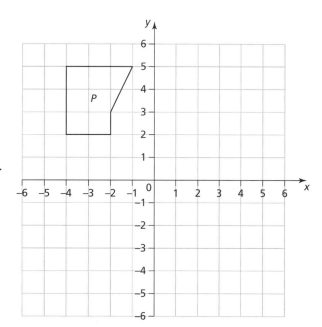

2 A shape *I* is shown on the grid.

a) Rotate shape *I* by 90° anticlockwise, centre (0, 0). Label the image *J*.

b) Reflect shape *J* in the *x*-axis. Label the image *K*.

c) Describe fully a single transformation that will map shape *I* to shape *K*.

....

....

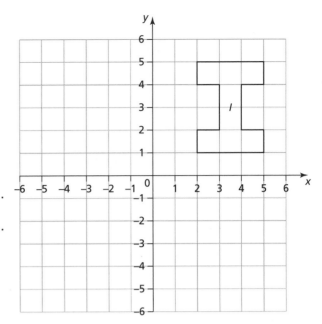

3 A shape *F* is shown on the grid.

a) Rotate *F* by 180°, centre (3, 0). Label the image *G*.

b) Translate shape *G* with vector $\begin{pmatrix} -6 \\ 0 \end{pmatrix}$. Label the image *H*.

c) Describe fully the single transformation that would map *F* directly onto *H*.

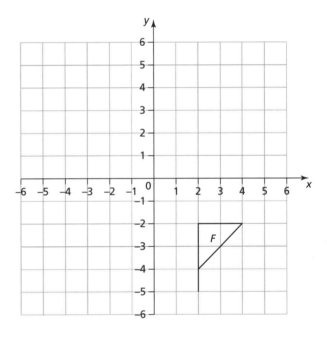

....

4 A shape *T* is shown on the grid.

a) Reflect shape *T* in the line *y* = *x*. Label the image *U*.

b) Rotate shape *U* by 90° clockwise, centre (0, 0). Label the image *V*.

c) Describe fully the single transformation that maps *T* directly onto *V*.

....

....

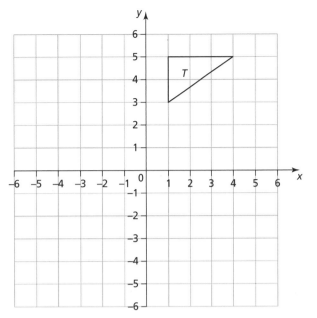

5 Use the grids to show that a rotation of 180°, centre (0, 0), followed by a translation with vector $\begin{pmatrix} 0 \\ 2 \end{pmatrix}$ is not the same as a translation with vector $\begin{pmatrix} 0 \\ 2 \end{pmatrix}$ followed by a rotation of 180°, centre (0, 0).

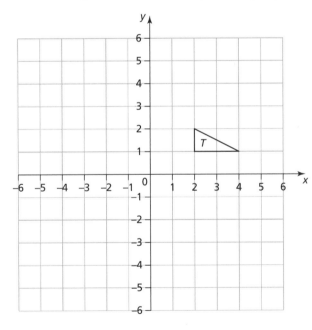

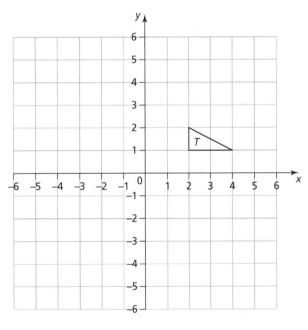

6 A shape, *K*, is reflected to shape *L*.
Shape *L* is translated to give shape *M*.

Give a reason why shapes *K* and *M* will be congruent.

....

....

Think about

7 A shape *T* is rotated by 90°, centre the origin.
The new shape, *U*, is then rotated by 180°, centre (2, 0), to give shape *V*.

Explore whether the following statement is true or false.

The single transformation that maps T directly to V is a rotation.

6.3 Enlargements

Summary of key points

To describe an enlargement, give the **scale factor** and the coordinates of the **centre of enlargement**.

If shape *Q* is an enlargement of shape *P* with a scale factor *k*, then:

- perimeter of $Q = k \times$ perimeter of *P*
- area of $Q = k^2 \times$ area of *P*

Exercise 3

1 Enlarge the quadrilateral by scale factor 2, centre (−1, 0).

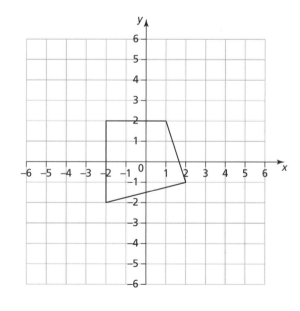

2 The diagram shows two triangles, *M* and *N*.

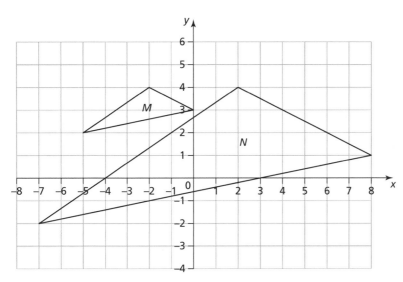

Describe fully the single transformation that maps *M* to *N*.

....

3 Describe fully each transformation from shape *P* to shape *Q*.

a)

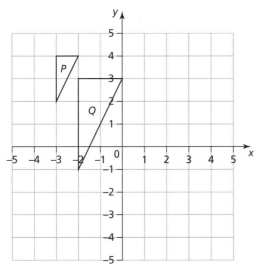

b)

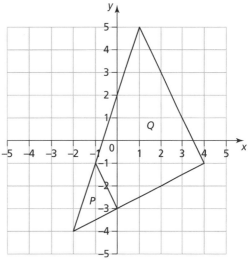

....

....

4 Describe fully each transformation from *Q* to *R*.

a)

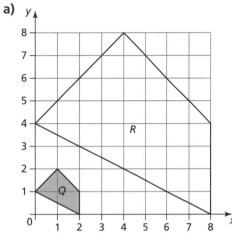

....

....

b)

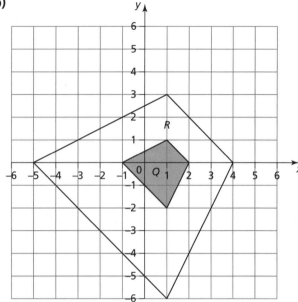

....

....

5 *PQRS* is a transformation of *ABCD*.

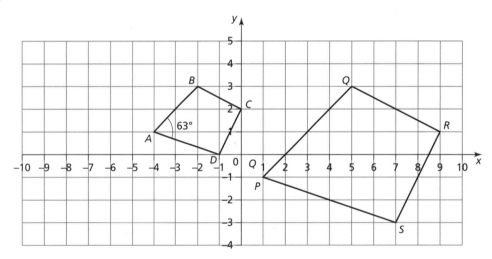

a) Describe fully the transformation that maps *ABCD* to *PQRS*.

....

b) Write down the size of angle *QPS*. ....°

c) Sabah says that *ABCD* and *PQRS* are congruent.

Is she correct?   Yes ☐   No ☐

Explain your answer.

....

6 A shape 3 cm wide has a perimeter of 10 cm and an area of 7 cm². The shape is enlarged with scale factor 2.

a) Find the perimeter of the enlarged shape.

.... cm

b) What is the area of the enlarged shape?

.... cm²

7 Philippa draws a polygon with a perimeter of 8 cm and an area of 4 cm². Beth draws an enlargement of Philippa's polygon with scale factor 10.

a) Beth says that her polygon will have a perimeter of 80 cm.

Is she correct?   Yes ☐   No ☐

Give a reason for your answer.

....

b) Beth also says that the enlarged shape will have an area of 40 cm².

Is she correct?   Yes ☐   No ☐

Explain your answer.

....

8 Freddie draws a rectangle measuring 8 cm by 2 cm.

He enlarges his rectangle on a photocopier. His enlarged rectangle has a perimeter of 1 m.

Find the scale factor of the enlargement.

....

9 A triangle has sides measuring 3 cm, 4 cm and 5 cm.

The triangle is enlarged. One of the sides of the image has length 60 cm.

Write down all the possible values for the perimeter of the enlarged triangle.

.... cm or .... cm or .... cm

7

Presenting and interpreting data 1

You will practice how to:

- Record, organise and represent categorical, discrete and continuous data. Choose and explain which representation to use in a given situation:
  - frequency polygons
  - stem-and-leaf and back-to-back stem-and-leaf diagrams.
- Interpret data, identifying patterns, trends and relationships, within and between data sets, to answer statistical questions. Make informal inferences and generalisations, identifying wrong or misleading information.

7.1 Frequency polygons

Summary of key points

Frequency polygons allow two sets of data to be compared on the same graph.

To plot a frequency polygon of grouped data, plot the frequency at the **midpoint** of each group.

A frequency diagram (dotted bars) and a frequency polygon (solid lines) representing the lengths in centimetres of a sample of 20 flowers are both drawn on the same axes.

To calculate the **midpoint** of a class interval, add the upper and lower boundaries of the interval and divide by 2.

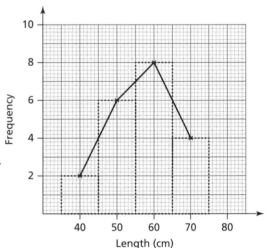

Think about

Collect data from your class, such as how many siblings each class member has or how many pets each class member has, and draw a frequency polygon to show the results.

1 The frequency table shows information about the times taken in seconds for a group of students to run 100 m.

Time, t (seconds)	Midpoint	Frequency
$10 \le t < 14$		1
$14 \le t < 18$		5
$18 \le t < 22$		13
$22 \le t < 26$		6

a) Write down the modal class interval.

....

b) Complete the table and draw a frequency polygon to show the information.

2 The table shows the average times taken for 40 teachers to travel to school.

Time, t (minutes)	Midpoint	Frequency
$5 \le t < 13$		4
$13 \le t < 21$		13
		15
$29 \le t < 37$		6
$37 \le t < 45$		

a) Complete the table.

b) Draw a frequency polygon to show the data.

3 The table shows the heights of a group of 35 boys and 35 girls.

Height, h (cm)	Midpoint	Frequency, boys	Frequency, girls
$140 \le h < 150$		0	6
		8	12
$160 \le h < 170$			15
		14	
$180 \le h < 190$		3	0

a) Complete the table.

b) Draw two frequency polygons to show this data.

c) Compare the heights of the boys and the girls.

....

....

....

4 The table shows the annual incomes of a group of employees of a technology company.

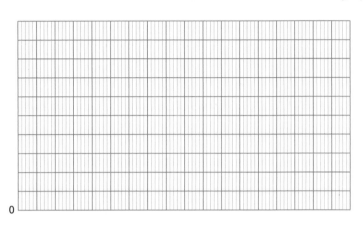

Income, I ($ thousands)	Frequency
10 < I ≤ 25	3
25 < I ≤ 40	22
40 < I ≤ 55	19
55 < I ≤ 70	12
70 < I ≤ 85	5

a) Draw a frequency polygon to show this information.

b) Find how many employees earn more than $70 000.

....

c) An employee is selected at random.

Find the probability that they earn $40 000 or less. Give your answer correct to 2 decimal places.

....

5 The frequency polygon shows the lengths of a sample of insects measured in a study.

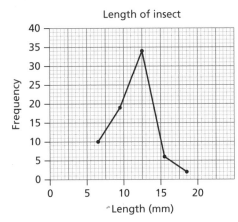

Length of insect

a) Use the frequency polygon to complete the table. The first row has been done for you.

Length, l (mm)	Midpoint	Frequency
$5 < l \le 8$	6.5	10

b) Find how many insects were measured in the sample.

....

c) Write down the modal class interval.

....

d) Manuel says that more than 10% of the insects are more than 14 mm in length.

Is Manuel correct? Explain your answer.

....

....

6 The frequency polygons show the resting heart rates in a sample of athletes and non-athletes.

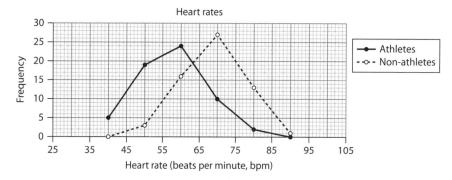

Heart rates

Are these statements true or false?

	True	False
The modal class interval for the athletes is 55 to 65 bpm.	☐	☐
Exactly 19 athletes have a resting heart rate of 50 bpm.	☐	☐
There are no non-athletes with a resting heart rate of 45 bpm or less.	☐	☐
120 people had their resting heart rates measured in the sample.	☐	☐
The modal class interval for the non-athletes is 75 to 85 bpm.	☐	☐

7.2 Back-to-back stem-and-leaf diagrams

Summary of key points

Two sets of data values can be compared by drawing a back-to-back stem-and-leaf diagram.

- There should be a common stem for both sets of data.
- The leaf values are written in order of size, with the smallest leaf values nearest the stem in each row.

Exercise 2

1 The back-to-back stem-and-leaf diagram shows the numbers of books read by 20 girls last year.

The numbers of books 20 boys read last year are listed below.

Boys		Girls
	0	8 9
	1	1 3 4 6
	2	4 5 6 7 9 9
	3	0 1 3 5 6
	4	1 4 5

Key 1 | 3 = 13 books
5 | 1 = 15 books

10	39	15	9	28	12	24	17	46	31
23	7	24	26	42	22	21	34	4	19

a) Add the data for the boys to the back-to-back stem-and-leaf diagram.

b) Find the total number of children who read at least 35 books.

....

c) The child who read the most books received a prize.

How many books did this child read?

....

d) Calculate the range for the number of books read by the girls.

....

2 The back-to back stem-and-leaf diagram shows the handspans of children aged 10 years and 12 years.

a) Find the difference between the longest handspan for the children aged 12 years and the longest handspan for the children aged 10 years.

.... cm

10 years		12 years
5 4 2	**14**	
9 6 6 4 3 0	**15**	
9 9 8 6 4 2 1	**16**	1 3 6
8 5 3 2	**17**	4 5 8 8 9
6 2	**18**	0 0 1 3 5 7 9
1	**19**	1 2 4 5 5 6
	20	2 4 5 9

Key 2 | **14** = 14.2 cm
16 | 1 = 16.1 cm

b) Complete the table.

Children aged 10 years	Children aged 12 years
Number of children = ...	Number of children = ...
Median handspan = ... cm	Median handspan = ... cm

3 The back-to-back stem-and-leaf diagram shows the weekly sales of computers in two shops for each of the last 25 weeks.

Shop 1		Shop 2
9 9 7 6 4	**0**	8 9
9 9 8 6 5 4 3 2 2 0	**1**	1 3 4 7 7 8 8
7 6 2 1	**2**	0 1 3 3 4 6 9 9
6 3	**3**	1 1 2 3 5 7
5 5 3 0	**4**	1 2

Key: 1 | **2** = 21 computer sales
2 | 0 = 20 computer sales

a) Both shops have a sales target of 36 computers per week. Staff in the store that meets this target most often during the 25-week period receive a bonus.

Which shop's staff receive a bonus?

....

b) Calculate the median number of computers sold in each shop.

Shop 1 ....   Shop 2 ....

Compare the average sales in the two shops.

....

c) Calculate the range for the number of computers sold in each shop.

Shop 1 ....   Shop 2 ....

Which shop had greater variability in the number of computers sold?

....

4 The back-to-back stem-and-leaf diagram shows the results for boys and girls in a maths assessment.

Boys		Girls
6 5	**0**	
9 7 4 3	**1**	5 8
6 5 4 3 1	**2**	3 5 9
2 0	**3**	2 4 6 9
2	**4**	1 4 8
	5	0

Key 3 | 1 = 13 marks
    4 | 1 = 41 marks

a) How many girls are in the class?

....

b) Write down the median of the girls' results.

....

c) Write down the range of the boys' results.

....

d) Write down the highest mark in the class.

....

e) Another boy takes the test the following day and scores 22 marks. Peter says this will increase the median value of the boys' results.

Is Peter correct? Give a reason for your answer.

....

....

5 The back-to-back stem-and-leaf diagram compares the numbers of words in sentences of two newspapers.

Daily News		Global Gazette
4 5	**0**	6 7 8
3 1 0	**1**	4 7
3 4 4 7	**2**	2 7 9
0 2	**3**	1 5
1	**4**	2 3

Explain what mistakes have been made.

....

....

8 Rounding and decimals

You will practice how to:

- Understand that when a number is rounded there are upper and lower limits for the original number.
- Estimate, multiply and divide decimals by integers and decimals.

8.1 Upper and lower limits

Summary of key points

When a quantity has been rounded or measured to a number of places, you can write an inequality showing the **upper and lower limits** on the quantity's true value.

If $x = 60$ to the nearest 10, then:	If $x = 5$ to the nearest whole number, then:
55 is the lowest number that rounds to 60	4.5 is the lowest number that rounds to 5
65 is the lowest number that does *not* round to 60	5.5 is the lowest number that does *not* round to 5
$55 \leq x < 65$	$4.5 \leq x < 5.5$

Exercise 1

1. Write an inequality showing the limits on each measurement. (You do not have to write the unit in the inequality.)

   a) A length, l, of 1200 m measured to the nearest 100 metres. ....

   b) A mass, m, of 88 g measured to the nearest gram. ....

   c) A number of people, n, of 90 to the nearest ten. ....

   d) A volume, V, of 4.1 ml measured to the nearest tenth of a millilitre. ....

   e) A time, t, of 8.46 s measured to the nearest 0.01 s. ....

2. The mass of a jackfruit, m kg, is measured. It is 2.7 kg rounded to 1 decimal place.

   Draw a ring around the numbers that are possible values of m.

   2.00   2.60   2.65   2.69   2.694   2.70

      2.71   2.75   2.758   2.76   2.85

3 The table shows six limits on different quantities, *a* to *f*. Write the rounded value of each quantity. State whether it has been rounded to the nearest 0.01, 0.1, 1, 10, 50, 100 or 1000.

The first row has been filled in for you.

Rounded value	Rounded to the nearest	Limits
5.32	0.01	$5.315 \leq a < 5.325$
		$78.5 \leq b < 79.5$
		$3950 \leq c < 4050$
		$8375 \leq d < 8425$
		$52.65 \leq e < 52.75$
		$1500 \leq f < 2500$

4 Complete the statements.

a) Rounding 7.58 to 1 decimal place is equivalent to rounding to .... significant figure(s).

b) Rounding 8200 to the nearest thousand is equivalent to rounding to .... significant figure(s).

c) Rounding 1822 to the nearest ten is equivalent to rounding to .... significant figure(s).

d) Rounding 867.3 to the nearest whole number is equivalent to rounding to .... significant figure(s).

5 Write an inequality showing the limits on each value of *x*.

a) *x* is 1500 after rounding to 3 significant figures. ....

b) *x* is 9000 after rounding to 1 significant figure. ....

c) *x* is 30 after rounding to 1 significant figure. ....

d) *x* is 0.20 after rounding to 2 significant figures. ....

6 Ebony and Johann work in an office building. They find out that the height of the building is approximately 120 m.

Ebony says, 'The height may have been rounded to the nearest whole metre.'

Johann says, 'That is incorrect. The height must have been rounded to the nearest ten, because the last digit is zero.'

Who is correct? Explain your answer.

....

....

7 If the height had been rounded to the nearest tenth of a metre in question 6, how would you expect it to be written differently?

8.2 Multiplying and dividing with integers and decimals

Summary of key points

Multiplying decimals

Example: 13.6 × 2.8

First estimate the result: 10 × 3 = 30

Multiply 13.6 × 2.8, ignoring the decimal points.

```
      1  3  6
×        2  8
  1  0  8  8
  2  7  2  0
  3  8  0  8
```

Then put the decimal point back in.

13.6 × 2.8 is equivalent to (136 × 28) ÷ 100, so the answer is 38.08.

Dividing decimals

Example: 5.85 ÷ 1.3

First estimate the result: 6 ÷ 1 ≈ 6

Find an equivalent calculation where the divisor is an integer:

5.85 ÷ 1.3 = 58.5 ÷ 13

Then carry out the division.

```
          4  .  5
13 ⟌ 5  8  .  ⁶5
```

Remember to keep the decimal points aligned.

So the answer is 4.5.

Exercise 2

1 Estimate and then calculate:

a) 63 × 2.7

b) 21.6 × 43

c) 7.4 × (−3.8)

d) 82.7 × 0.76

....

....

2 **Estimate and then calculate:**

a) 364 ÷ 1.4

b) −496 ÷ (−3.1)

....

....

c) 1.92 ÷ 0.12

d) (−0.756) ÷ 2.1

....

....

3 **Fabric costs $2.45 per metre. Alex buys 3.6 metres of fabric.**

a) Estimate the total cost of Alex's fabric.

$.....................

b) Calculate the total cost of Alex's fabric.

$......................

4 A box of fertiliser has these instructions.

Use 0.035 kg of fertiliser per
square metre of grass

Estimate and then calculate how many square metres of grass can be fertilised
using one full box of fertiliser. A full box of fertiliser contains 1.96 kg.

.... square metres

5 Complete this multiplication grid.

×	7.6	....
0.54	....	0.918
4.9	....	....

6 Find the value of 2.8^3.

....

7 If ■ = 14.4 × 3.6 and ▲ = 2.52 ÷ 1.4, calculate the value of $\frac{■}{▲}$.

....

8 Estimate and then calculate:

a) 3 × 3.7 × (−7)

b) (−2.2) × (−3) × (−8)

....

....

c) (3.7 + 1.32) × 2.9

d) 5.6 × (3.22 ÷ 1.4)

....

....

9 Elliot wants to find 3.78 ÷ 0.14. His working is below.

$$\frac{3.78}{0.14} = \frac{378}{14}$$

$$\begin{array}{r} 2\ 7 \\ 14\overline{)37^98} \end{array}$$

378 ÷ 14 = 27, so 3.78 ÷ 0.14 = 0.27

Describe the mistake he has made. Write the correct answer.

....

....

Summary of key points

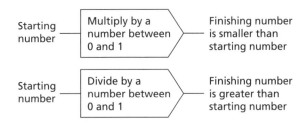

| Starting number | → | Multiply by a number between 0 and 1 | → | Finishing number is smaller than starting number |

| Starting number | → | Divide by a number between 0 and 1 | → | Finishing number is greater than starting number |

Exercise 3

1 In each part, decide whether the answer is more than or less than 50.

    a) 50 × 1.1 ....

    b) 50 ÷ 0.44 ....

    c) 50 ÷ 1.04 ....

    d) 50 × 0.86 ....

2 Insert the correct sign (< or >) to make each statement true.

    a) 4.8 × 0.7 .... 4.8   **b)** 36 × 1.25 36   **c)** 98 ÷ 2.06 98

    d) 12.6 × 0.88 .... 12.6   **e)** 3.4 ÷ 0.06 .... 3.4   **f)** $14 \div \frac{2}{3}$ 14

3 Insert the correct operation (× or ÷) to make each statement true.

    a) 7.4 .... 0.75 = 5.55   **b)** 1.8 .... 0.4 = 4.5

    c) 4.5 .... 0.25 = 18   **d)** 75 .... 0.34 = 25.5

4 Write a positive number to complete each statement.

    a) 32 ÷ .... > 32   **b)** 0.48 × .... = 0.48

    c) 18 ÷ .... < 18   **d)** .... × 0.94 < 0.94

5 Explain how you know each of these calculations is incorrect.

    a) 48 × 0.985 = 49.28

    ....

    b) 70 ÷ 1.05 = 73.5

    ....

    c) 64 ÷ 0.8 = 60

    ....

9 Functions and formulae

You will practice how to:

- Understand that a function is a relationship where each input has a single output.
- Generate outputs from a given function and identify inputs from a given output by considering inverse operations (including indices).
- Understand that a situation can be represented either in words or as a formula (including squares and cubes), and manipulate using knowledge of inverse operations to change the subject of a formula.

9.1 Functions

Summary of key points

A **function** can be represented by a function machine, a mapping diagram or an algebraic statement.

The function machine for $x \to (x + 5)^2$ is:

$$x \longrightarrow \boxed{+\ 5} \longrightarrow \boxed{\text{square}} \longrightarrow (x + 5)^2$$

The **inverse function** can be found by reversing the function process:

$$\sqrt{x} - 5 \longleftarrow \boxed{-\ 5} \longleftarrow \boxed{\text{square root}} \longleftarrow x$$

So the inverse function of $x \to (x + 5)^2$ is $x \to \sqrt{x} - 5$.

Exercise 1

1 Match each function machine with the correct algebraic expression.

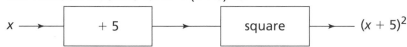

a)
$$x \longrightarrow \boxed{\times\ 3} \longrightarrow \boxed{+\ 4} \longrightarrow \qquad\qquad x \to 3(x + 4)$$

b)
$$x \longrightarrow \boxed{+\ 4} \longrightarrow \boxed{\times\ 3} \longrightarrow \qquad\qquad x \to 4\sqrt{x}$$

c)
$$x \longrightarrow \boxed{\text{square}} \longrightarrow \boxed{-\ 3} \longrightarrow \qquad\qquad (x - 3)^2$$

d)
$$x \longrightarrow \boxed{-\ 3} \longrightarrow \boxed{\text{square}} \longrightarrow \qquad\qquad x \to \sqrt{x - 4}$$

e)
$$x \longrightarrow \boxed{\text{square root}} \longrightarrow \boxed{\times\ 4} \longrightarrow \qquad\qquad x \to 3x + 4$$

f)
$$x \longrightarrow \boxed{-\ 4} \longrightarrow \boxed{\text{square root}} \longrightarrow \qquad\qquad x \to x^2 - 3$$

2 Complete the input and output tables for each function.

a) $x \to x^2 + 3$

input	output
−3	
0	
	4
5	
	52

b) $x \to (x + 2)^2$

input	output
−3	
1	
	25
	64
10	

c) $x \to \dfrac{12}{x}$

input	output
−3	
1	
2	
	4
12	

3 For each function, choose the correct inverse from the box.

> $x \to 4x$ $x \to x - 4$
>
> $x \to \dfrac{x}{4} - 1$ $x \to \dfrac{x}{4}$
>
> $x \to x + 4$ $x \to \dfrac{x - 1}{4}$

a) $x \to 4x$ $x \to$ ....

b) $x \to x - 4$ $x \to$ ....

c) $x \to x + 4$ $x \to$ ....

d) $x \to 4x + 1$ $x \to$ ....

4 Complete the inverse functions.

Function	Inverse function
a) $x \to 5x - 1$	$x \to \dfrac{x\,\square}{5}$
b) $x \to 7x - 5$	$x \to \dfrac{x + 5}{\square}$
c) $x \to \dfrac{x}{2} + 1$	$x \to \square (x - 1)$
d) $x \to \dfrac{x - 8}{3}$	$x \to \square$

5 Find the function for each mapping diagram.

a)

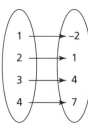

....

b)

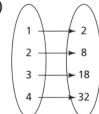

....

c)

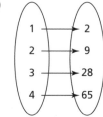

....

6 Complete the mapping diagram for each function.

a) $x \to \dfrac{24}{x} + 3$

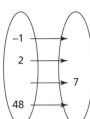

b) $x \to 60 - x^2$

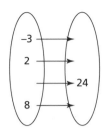

c) $x \to \sqrt{\dfrac{x+6}{2}}$

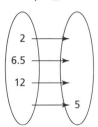

7 Match each function with its inverse function.

Draw a ring around the odd one out.

$x \to 3x + 7$	$x \to x^3 + 7$	$x \to (x - 2)^2$	$x \to x^2 + 2$

$x \to \sqrt{x} + 2$	$x \to \sqrt{x - 2}$	$x \to 7x - 3$	$x \to \sqrt{\dfrac{x}{2}}$

$x \to \dfrac{x + 3}{7}$	$x \to 2x^2$	$x \to \dfrac{x - 7}{3}$

8 Carole is setting up a language translation business.
She will charge a $20 fee plus $2 for every 100 words translated.

a) Write a function machine to calculate the charge for *w* words to be translated.

b) Enrico pays $88 for a document to be translated.

Find how many words were in the document.

....

A self-inverse function is one where the inverse function is the same as the original function.

Michael thinks that any function of the form $x \rightarrow \dfrac{a}{x}$, where a is a number, will be a self-inverse function.

Set up a spreadsheet to try the function for different values of a and different values of x.

Is Michael correct?

9.2 Formulae

Summary of key points

A shape is made by joining together a rectangle and a square.

The area of the rectangle is $2a \times b = 2ab$ cm².

The area of the square is $a \times a = a^2$ cm².

The formula for the total area, A cm², of the shape is $A = 2ab + a^2$.

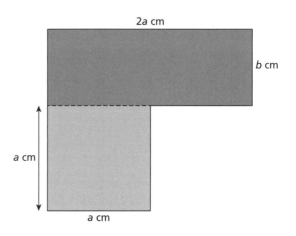

In the formula $y = mx + c$, y is the **subject** of the formula.

You can rearrange a formula to make a different variable the subject.

For example, for $y = mx + c$:

To make c the subject:

$$y = mx + c$$
$$-mx \downarrow \qquad \downarrow -mx$$
$$y - mx = c$$

So, the formula is:

$c = y - mx$.

To make m the subject:

$$y = mx + c$$
$$-c \downarrow \qquad \downarrow -c$$
$$y - c = mx$$
$$\div x \downarrow \qquad \downarrow \div x$$
$$\frac{y - c}{x} = m$$

So, the formula is $m = \dfrac{y - c}{x}$.

1 **Flynn finds these instructions on the internet for finding the area of a rhombus.**

Step 1: Multiply together the lengths of the two diagonals.

Step 2: Divide by 2.

Find a formula for the area, A cm², of a rhombus with diagonals of length t cm and u cm.

....

2 **The diagram shows an isosceles triangle.**

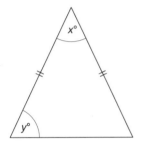

Find a formula for y in terms of x. ....

3 **The cost of a holiday is $\$A$ for each adult and $\$C$ for each child.**

The total cost of the holiday can be paid in six equal monthly amounts.

Find a formula for the amount, $\$M$, paid each month for a holiday for 2 adults and 3 children.

....

4 **Tomas builds a patio.**

He uses p large square patio stones and q small square patio stones.

The large patio stones have a length of x m.

The small patio stones have a length of y m.

Find a formula for the total area, A m², of the stones he uses.

....

5 *P* is the product of two consecutive numbers, *n* and *n* + 1.

Write a formula for *P*. Give your answer in a form without brackets.

....

6 The diagram shows a pentagon formed from a rectangle and a triangle.

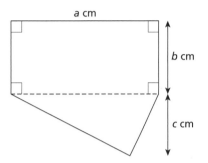

Find a formula for the total area, *A* cm², of the shape. ....

7 The diagram shows an object made from a cube and a cuboid.

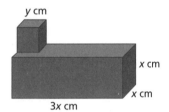

Find a formula for the volume, *V* cm³, in terms of *x* and *y*. ....

Think about

8 The area of a shape is given by the formula $A = x^2 - \frac{1}{2}xy$.

Draw a shape that this formula could be used for. Label the sides with their dimensions in *x* and *y*. Is the shape you have found the only one that could have this area formula?

9 The formula for the volume of a square-based prism is $V = x^2h$.

Find:

a) V when $x = 24$ and $h = 1.25$

$V = \ldots\ldots\ldots$

b) h when $V = 175$ and $x = 5$.

$h = \ldots\ldots\ldots$

10 Match the pairs of equivalent formulae.

$$y = \frac{3(x - 6)}{5}$$

$$y = \frac{3x - 6}{5}$$

$$y = \frac{3x}{5} - 6$$

$$y = 3(5x - 6)$$

$$x = \frac{5y + 6}{3}$$

$$x = \frac{y + 18}{15}$$

$$x = \frac{5y}{3} + 6$$

$$x = \frac{5(y + 6)}{3}$$

11 Make x the subject of each formula.

a) $y = \frac{2x + 7}{4}$

b) $y = \frac{9x}{2} + 11$

$x = \ldots\ldots\ldots$

$x = \ldots\ldots\ldots$

c) $y = \frac{5(x - 1)}{4}$

d) $y = \frac{2(3x + 2)}{7}$

$x = \ldots\ldots\ldots$

$x = \ldots\ldots\ldots$

12 $R = \dfrac{V}{I}$ is a formula that relates to electricity.

Rearrange the formula to make:

a) V the subject

b) I the subject.

$V =$ ………..

$I =$ ………..

13 Jean-Paul is trying to make p the subject of the formula $q = p^2 + 8$. Here is his working:

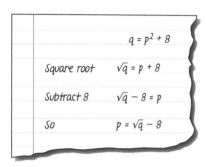

$q = p^2 + 8$

Square root $\sqrt{q} = p + 8$

Subtract 8 $\sqrt{q} - 8 = p$

So $p = \sqrt{q} - 8$

What mistake has Jean-Paul made?

....

....

14 Make y the subject of each formula.

a) $V = x^2y$

b) $T = \dfrac{3y + a}{k}$

$y =$ ………..

$y =$ ………..

c) $h = 4ay - 5g$

d) $A = \dfrac{1}{2}xy$

$y =$ ………..

$y =$ ………..

e) $m = y^2 - 5$

f) $d = \sqrt{\dfrac{y+2}{5}}$

$y = \ldots\ldots\ldots\ldots\ldots$

$y = \ldots\ldots\ldots\ldots\ldots$

g) $k = (y-1)^2$

h) $z = \sqrt{2y} - 3$

$y = \ldots\ldots\ldots\ldots\ldots$

$y = \ldots\ldots\ldots\ldots\ldots$

15 The formula for the surface area, *S*, of a cylinder is $S = 2\pi r^2 + 2\pi rh$.

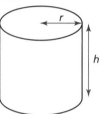

a) Rearrange the formula to make *h* the subject.

$\ldots\ldots\ldots\ldots\ldots$

b) Is it possible to rearrange the formula to make *r* the subject?

$\ldots\ldots\ldots\ldots\ldots$

Explain how you know.

$\ldots\ldots\ldots\ldots\ldots\ldots\ldots\ldots\ldots\ldots\ldots\ldots\ldots\ldots\ldots\ldots\ldots\ldots\ldots\ldots\ldots\ldots\ldots\ldots\ldots\ldots\ldots\ldots\ldots\ldots\ldots\ldots\ldots\ldots$

$\ldots\ldots\ldots\ldots\ldots\ldots\ldots\ldots\ldots\ldots\ldots\ldots\ldots\ldots\ldots\ldots\ldots\ldots\ldots\ldots\ldots\ldots\ldots\ldots\ldots\ldots\ldots\ldots\ldots\ldots\ldots\ldots\ldots\ldots$

10 Accurate drawing

You will practice how to:

- Construct 60°, 45° and 30° angles and regular polygons.
- Use knowledge of bearings and scaling to interpret position on maps and plans.
- Use knowledge of coordinates to find points on a line segment.

10.1 Construction

Summary of key points

Construct means to draw accurately.

To construct an angle of 60°:

- Construct an equilateral triangle.

To construct an angle of 90°:

- Construct the perpendicular bisector of a line segment.

To construct an angle of 30°:

- Bisect an angle of 60°.

To construct an angle of 45°:

- Bisect an angle of 90°.

Inscribing a hexagon and an equilateral triangle inside a circle

- Set the compasses to match the radius of the circle. Then mark divisions on the circumference of the circle.
- Join adjacent divisions to form a regular hexagon.
- Join every other division to form an equilateral triangle.

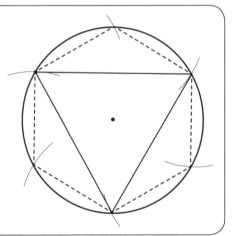

A polygon is **inscribed** inside a circle if all vertices of the polygon lie on the circumference of the circle.

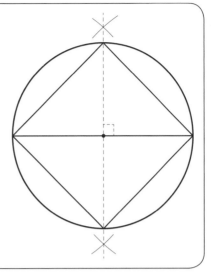

Inscribing a square inside a circle

- Draw a diameter and bisect it to form a perpendicular diameter.
- Join the ends of these diameters together to form a square.

To inscribe an octagon inside a circle:

- Follow the first step for inscribing a square.
- Bisect two adjacent 90° angles at the centre of the circle and mark the points where the bisectors intersect the circle.
- Join together the points on the circumference.

Exercise 1

1. **Complete this construction of a 60° angle.**

2 Complete this construction of a 45° angle.

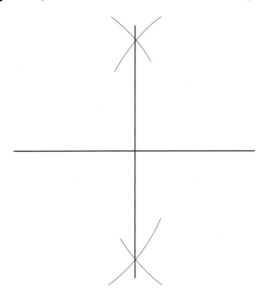

3 Complete this construction to inscribe a square within the circle.

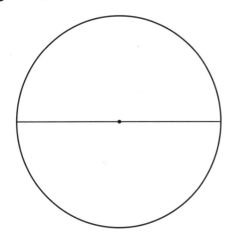

4 Complete this construction to inscribe an octagon within the circle.

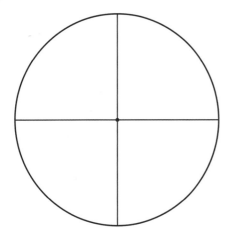

5 Use a straight edge and compasses to inscribe an equilateral triangle inside this circle.

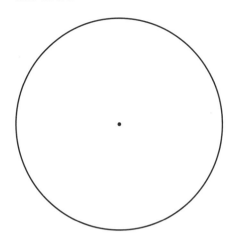

6 Freda is trying to inscribe a regular hexagon inside a circle.

This is the start of her construction.

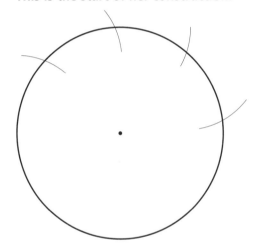

What mistake has Freda made?

....

....

7 **Construct the logo shown on the right.**

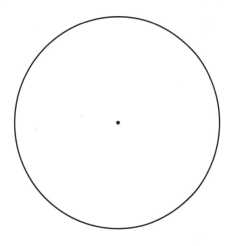

8 **Construct this triangle.**

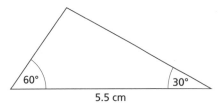

60° 30°

5.5 cm

10.2 Bearings

Summary of key points

Bearings are angles measured clockwise from north. They measure direction and are written with three digits. For example, 090° is due east.

When measuring or drawing a bearing, remember to begin by drawing a **north line** if it isn't already given.

Maps are a type of scale drawing.

The scale on the map is given as a ratio. The scale 1 : 100 000 means that 1 cm on the map represents an actual distance of 100 000 cm, which is 1000 m, which is 1 km.

1 Max leaves his car in a car park (*P*) and goes for a walk.

He walks 2 km on a bearing of 080°.

He turns and walks 1.5 km on a bearing of 165°.

Then he walks 2.5 km on a bearing of 280°.

a) Draw a scale drawing to show Max's walk. Use a scale of 1 : 50 000.

N

P

b) Max then walks straight back to his car.
How far does Max have to walk?

.... km

2 The map shows the positions of two villages, *A* and *B*. It is drawn using a scale of 1 : 250 000.

N

A

N

B

a) Find the actual distance (in kilometres) between village *A* and village *B*.

.... km

b) Measure the bearing of village *B* from village *A*.

.... °

c) Village *C* is on a bearing of 270° from village *B*. The distance between village *B* and village *C* is 20 km. Mark the position of village *C* on the diagram and find the actual distance between village *C* and village *A*.

.... km

d) Village *E* is on a bearing of 070° from village *A* and 330° from village *B*. Eva says that village *E* is 10 km from village *B*.

Is Eva correct?   Yes ☐   No ☐

Explain your answer.

....

....

3 **The diagram is a scale drawing showing part of a theme park.**

N

log flume •

rollercoaster •   scale 1 cm : 50 m

A carousel is on a bearing of 120° from the log flume and 055° from the rollercoaster.

a) By marking the position of the carousel on the scale drawing, find the actual distance between the carousel and the rollercoaster.

.... m

b) A big wheel is 140 m due west of the carousel.

Find the bearing of the big wheel from the rollercoaster.

.... °

10.3 Points on a line segment

Summary of key points

A **line segment** is a section of a line. It has two end points.

Example:

Find the point one quarter of the way along the line segment from (−7, 13) to (9, −7).

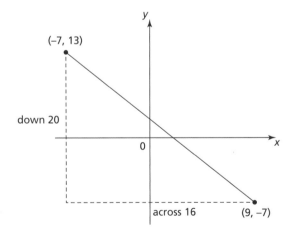 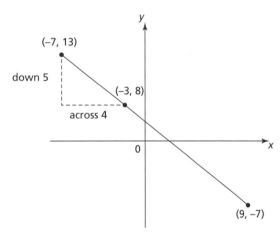

The point one quarter of the way along the line segment is (−3, 8).

1 **Write down the coordinates of the point:**

a) one third of the way along the line segment *AB*

b) one fifth of way along the line segment *AB*

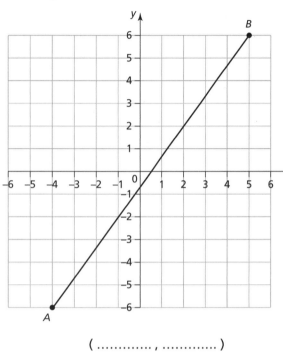

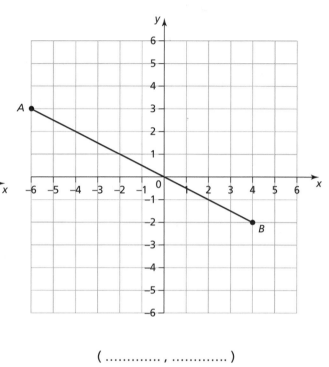

(............. , ...)

(............. , ...)

2 **a)** Find the point that lies one third of the way along the line segment *AB* joining *A*(−1, 5) and *B*(11, −1).

(............. , ...)

b) Find the point that lies one fifth of the way along the line segment *PQ* joining *P*(−3, −7) and *Q*(17, 8).

(............. , ...)

c) Find the point that lies one quarter of the way along the line segment *MN* joining *M*(10, 9) and *N*(−14, −7).

(............. , ...)

d) Find the point that lies two thirds of the way along the line segment *EF* joining *E*(0, 4) and *F*(18, 16).

(............. , ...)

3 *P* has coordinates (4, −2). *Q* has coordinates (−1, −8). *Q* is one quarter of the way along line segment *PR*.

Find the coordinates of *R*.

(............. , ...)

4 A triangle has coordinates *A*(6, 8), *B*(3, −4) and *C*(9, −4).

a) *D* is the point one third of the way along line segment *AB*.

E is the midpoint of line segment *BC*.

F is the point one third of the way along line segment *AC*.

Find the coordinates of *D*, *E* and *F*.

D(.......... , ...) *E*(.......... , ...) *F*(.......... , ...)

b) Andrew thinks that triangle *DEF* is isosceles.

Is Andrew correct?   Yes ☐   No ☐

Explain your answer.

....

....

5 *AB* is a line segment where *A* has coordinates (*a*, *b*) and *B* has coordinates (8, 20). *P* is the point one third of the way along *AB*. The coordinates of *P* are (2*a*, 2*b*).

Find the coordinates of *A*.

(............. , ...)

Find five line segments *AB* which have the following property:

The coordinates of the point one third of the way from *A* to *B* are (1, −3).

11 Fractions

You will practice how to:

- Deduce whether fractions will have recurring or terminating decimal equivalents.
- Estimate, add and subtract proper and improper fractions, and mixed numbers, using the order of operations.
- Estimate, multiply and divide fractions, interpret division as a multiplicative inverse, and cancel common factors before multiplying or dividing.

11.1 Recurring and terminating decimals

Summary of key points

To check whether a fraction has a **terminating** or **recurring decimal** equivalent, look at the prime factors of its denominator. The fraction must be in its simplest form.

- If all of the prime factors are 2s and/or 5s, the decimal terminates.
- If there are any other prime factors, the decimal recurs.

Examples:

$\frac{3}{40}$	$\frac{31}{70}$
$40 = 2^3 \times 5$	$70 = 2 \times 5 \times 7$
All of the prime factors are 2 or 5, so $\frac{3}{40}$ has a terminating decimal equivalent.	Not all of the prime factors are 2 or 5, so $\frac{31}{70}$ has a recurring decimal equivalent.

Exercise 1   1–3

1 Predict whether each fraction has a recurring (R) or terminating (T) decimal equivalent, by marking the R or T box.

Then check your answers using division.

a) $\frac{2}{9}$ R ☐ T ☐   b) $\frac{5}{8}$ R ☐ T ☐

   Division: Division:

c) $\frac{7}{15}$ R ☐ T ☐   d) $\frac{11}{25}$ R ☐ T ☐

   Division: Division:

2 Write each fraction its simplest form.

Without calculating, state whether its decimal equivalent terminates or recurs.

Fraction	Fraction in simplest form	Terminating/recurring decimal equivalent?
$\frac{9}{24}$		
$\frac{4}{6}$		
$\frac{7}{49}$		
$\frac{12}{15}$		

3 Devi says, 'The only fractions with terminating decimal equivalents are fractions with denominators 2^n and 5^n, where n is a whole number.'

Is she correct? Explain your answer.

....

....

4 **a)** Using a calculator, complete the table showing the decimal equivalents of elevenths from $\frac{1}{11}$ to $\frac{7}{11}$. Write these using dot notation.

Fraction	$\frac{1}{11}$	$\frac{2}{11}$	$\frac{3}{11}$	$\frac{4}{11}$	$\frac{5}{11}$	$\frac{6}{11}$	$\frac{7}{11}$
Decimal equivalent							

b) Predict the decimal equivalents of $\frac{8}{11}$, $\frac{9}{11}$ and $\frac{10}{11}$. Use a calculator to check your answers.

$\frac{8}{11} =$ $\frac{9}{11} =$ $\frac{10}{11} =$

....

c) Explain why none of the elevenths have a terminating decimal equivalent.

....

Think about

5 How can you use your answer to question 4 to show that $0.\dot{9} = 1$?

Summary of key points

In addition and subtraction of fractions, follow the usual order of operations. Below are two methods of finding $4\frac{3}{4} - \left(\frac{19}{16} + 1\frac{5}{8}\right)$. Do the calculation in brackets first.

Writing as mixed numbers first:	Writing as improper fractions first:
$4\frac{3}{4} - \left(\frac{19}{16} + 1\frac{5}{8}\right) = 4\frac{3}{4} - \left(1\frac{3}{16} + 1\frac{5}{8}\right)$	$4\frac{3}{4} - \left(\frac{19}{16} + 1\frac{5}{8}\right) = \frac{19}{4} - \left(\frac{19}{16} + \frac{13}{8}\right)$
$= 4\frac{12}{16} - \left(1\frac{3}{16} + 1\frac{10}{16}\right)$	$= \frac{76}{16} - \left(\frac{19}{16} + \frac{26}{16}\right)$
$= 4\frac{12}{16} - 2\frac{13}{16}$	$= \frac{76}{16} - \frac{45}{16}$
$= 3\frac{28}{16} - 2\frac{13}{16}$	$= \frac{31}{16}$
$= 1\frac{15}{16}$	$= 1\frac{15}{16}$

Exercise 2

1 Estimate and then calculate, writing your answers as mixed numbers:

a) $4\frac{5}{8} - \frac{15}{4}$

b) $\frac{7}{3} + 2\frac{3}{4}$

....

....

c) $2\frac{1}{3} + 1\frac{7}{12} + \frac{5}{8}$

d) $3\frac{2}{5} - 1\frac{3}{8} + \frac{3}{4}$

....

....

2 Fill in the missing digits. Write the answers in their simplest form.

a) $\dfrac{14}{9} + 3\dfrac{1}{\square} = 1\dfrac{\square}{18} + 3\dfrac{3}{18}$

b) $4\dfrac{5}{6} + \dfrac{1}{8} - 2\dfrac{1}{3} = \square + \dfrac{20}{24} + \dfrac{3}{\square} - \square - \dfrac{\square}{24}$

$= \square\dfrac{\square}{\square}$

$= \square\dfrac{\square}{\square}$

3 Draw a ring around the expressions below that are equivalent to $2\dfrac{1}{8} - \dfrac{15}{16} + \dfrac{5}{2}$.

A $\quad 2\dfrac{2}{16} - \dfrac{15}{16} + \dfrac{40}{16}$

B $\quad \dfrac{3}{8} - \dfrac{15}{16} + \dfrac{5}{2}$

C $\quad \dfrac{17}{16} - \dfrac{15}{16} + \dfrac{21}{16}$

D $\quad \dfrac{17}{8} - \dfrac{15}{16} + \dfrac{5}{2}$

E $\quad 2\dfrac{2}{16} - \dfrac{15}{16} + 2\dfrac{8}{16}$

Think about

4 If you were to use one of the expressions in question 3 to start calculating
$2\dfrac{1}{8} - \dfrac{15}{16} + \dfrac{5}{2}$, which one would you choose and why?

5 Max wants to find $5\dfrac{1}{6} + \left(\dfrac{4}{5} + \dfrac{5}{6}\right)$.

He says, 'First I find $5\dfrac{1}{6} + \dfrac{5}{6} = 6$. Then I find $6 + \dfrac{4}{5} = 6\dfrac{4}{5}$.'

Is his method correct? Explain why.

....

....

6 Estimate and then calculate, writing your answers as mixed numbers:

a) $5 - \left(\dfrac{1}{8} + \dfrac{7}{12}\right)$

b) $2\dfrac{1}{3} + \left(1\dfrac{5}{6} - \dfrac{8}{9}\right)$

....

....

c) $4\dfrac{5}{6} - \left(3\dfrac{1}{4} + \dfrac{1}{8}\right)$

d) $\left(\dfrac{16}{3} - 1\dfrac{11}{30}\right) - \left(\dfrac{3}{5} + \dfrac{11}{15}\right)$

....

....

11.3 Multiplying and dividing fractions

Summary of key points

Multiplying two fractions	Dividing two fractions
(1) Change any mixed numbers to improper fractions.	(1) Change any mixed numbers to improper fractions.
(2) Cancel any common factors.	(2) Turn the second fraction upside down and multiply.
(3) Multiply the numerators and the denominators together.	Example:

Example:

$$3\frac{1}{3} \times 1\frac{3}{5} = \frac{\overset{2}{\cancel{10}}}{3} \times \frac{8}{\cancel{5}_1}$$

$$= \frac{16}{3}$$

$$= 5\frac{1}{3}$$

$$1\frac{4}{5} \div \frac{6}{7} = \frac{9}{5} \div \frac{6}{7}$$

$$= \frac{\overset{3}{\cancel{9}}}{5} \times \frac{7}{\cancel{6}_2}$$

$$= \frac{21}{10} \text{ or } 2\frac{1}{10}$$

Exercise 3

1 **Find the answers. Write answers greater than 1 as mixed numbers, and write fractions in their simplest form.**

a) $1\frac{1}{4} \times \frac{5}{8}$

b) $\frac{7}{6} \times \frac{5}{3}$

....

....

c) $2\frac{1}{6} \times 1\frac{3}{7}$

d) $\frac{14}{11} \times 1\frac{3}{7}$

....

....

2 A small cake is made using $\frac{3}{8}$ kg of flour. Luca wants to make a medium cake

that has a mass $1\frac{1}{2}$ times greater than the small cake.

Write in a fraction how much flour Lucas needs for his cake.

.... kg

3 Choose the correct answer for each of the division calculations from the options below.

A	B	C	D	E
$2\frac{1}{3}$	$\frac{9}{14}$	$\frac{1}{12}$	$\frac{5}{8}$	$1\frac{1}{3}$

a) $\frac{2}{3} \div 8$ ....

b) $\frac{7}{9} \div \frac{1}{3}$ ....

c) $\frac{3}{8} \div \frac{7}{12}$ ....

d) $\frac{5}{6} \div 1\frac{1}{3}$ ....

e) $1\frac{5}{7} \div 2\frac{2}{3}$ ....

f) $2\frac{4}{5} \div 2\frac{1}{10}$ ....

4 For each calculation, tick the correct box.

Calculation	Answer is larger than a	Answer is smaller than a
$1\frac{3}{5} \times a$	☐	☐
$a \div \frac{2}{3}$	☐	☐
$a \div \frac{7}{6}$	☐	☐
$a \times \frac{1}{4}$	☐	☐
$a \div 2\frac{2}{3}$	☐	☐
$\frac{9}{8} \times a$	☐	☐

5 Find the answers, giving each answer as a mixed number in its simplest form.

a) $3\frac{3}{4} \div \left(\frac{1}{4} + \frac{7}{8}\right)$

b) $5 - \left(\frac{5}{8} + 1\frac{3}{4}\right)$

....

....

c) $\frac{8}{5} + 1\frac{2}{3} \times \frac{3}{2}$

d) $8\frac{1}{6} - 4 \div \frac{6}{7}$

....

....

12 Probability 1

You will practice how to:

- Understand that the probability of multiple mutually exclusive events can be found by summation and all mutually exclusive events have a total probability of 1.
- Design and conduct chance experiments or simulations, using small and large numbers of trials. Calculate the expected frequency of occurrences and compare with observed outcomes.

12.1 Mutually exclusive events

Summary of key points

Outcomes are called **mutually exclusive** if they cannot occur at the same time.

The sum of the probabilities of all the mutually exclusive outcomes of an experiment is always 1 (as one of the outcomes must occur).

Example:

A bag contains balls that are coloured either red, yellow, green or blue. The probabilities are shown in the table.

Colour	Red	Yellow	Green	Blue
Probability	0.25	0.15	$2k$	k

The table shows all the possible outcomes. Because the outcomes are mutually exclusive, the sum of the probabilities is 1.

Therefore, $0.25 + 0.15 + 2k + k = 1$

$$0.4 + 3k = 1$$

So $k = 0.2$

Exercise 1

1 Sam throws a biased six-sided dice. The table shows some of the probabilities.

Score	1	2	3	4	5	6
Probability	0.2	0.1	0.25	0.1	0.15	

Work out the probability that the score on the dice is 6.

....

2 Cars at a junction can turn left, turn right or go straight on. The table shows the probabilities of two of these outcomes.

Direction	Left	Right	Straight on
Probability	0.35	0.44	

Find the probability that a car will go straight on.

....

3 A gardener puts some plants in his gardens. The plants can have white or pink or red flowers.

The probability that a plant has white flowers is $\frac{5}{12}$.

The probability that a plant has pink flowers is $\frac{1}{3}$.

a) What is the probability that a plant has red flowers?

....

b) Find the probability that the plant has flowers that are *not* white.

....

4 All the children in a school choose one extracurricular activity. They can choose swimming, art, singing or drama.

Two of the probabilities are shown in the table.

Activity	Swimming	Art	Singing	Drama
Probability	0.36	0.12		

The probability of a child choosing drama is twice the probability of choosing art.

a) Complete the table.

b) Find the probability that a child chooses swimming or singing.

....

5 A shop sells small, medium and large cakes.

The probability that a customer buys a small cake is 0.4.

A customer is three times more likely to buy a medium cake than a large cake.

a) Work out the probability that a customer buys a large cake.

....

b) Find the probability that a customer does not buy a medium cake.

....

6 Trains arriving at a station are early, on time or late.

The table shows the probability of each outcome.

Outcome	Early	On time	Late
Probability	0.42	$2a + 0.1$	a

a) Find the value of a.

....

b) Halim says that a train is more likely to arrive early than it is to arrive on time.

Is Halim correct? Yes ☐ No ☐

Show how you worked out your answer.

....

....

Think about

7 What is the advantage of using mutually exclusive questions in the design of a questionnaire?

12.2 Experimental probability

Summary of key points

Expected frequency of an event = theoretical probability of the event occurring × number of trials

Increasing the number of trials gives a more accurate estimate of theoretical probability.

Exercise 2

1 **a)** The probability that Bobby wins at chess is 0.8. In a tournament he plays 30 games.

How many games would he be expected to win?

....

b) Marta rolls a fair six-sided dice 240 times.

Calculate the expected frequency of it landing on 4.

....

2 **Jemma rolls a six-sided dice 60 times. It lands on 2 fifteen times.**

a) Find the expected frequency of the dice landing on a 2.

....

b) Is Jemma's dice biased? Explain your reasoning.

....

....

3 Freddy spins a spinner with three sectors: pink, grey and blue. He spins it 200 times and records the number of times it lands on pink after every 50 spins.

The table shows the relative frequency of the spinner landing on pink after each 50 spins.

Number of spins	50	100	150	200
Relative frequency of landing on pink	0.22	0.36	0.40	0.42

a) Find how many times the spinner landed on the pink sector after 100 spins.

....

b) Use the table to complete the graph.

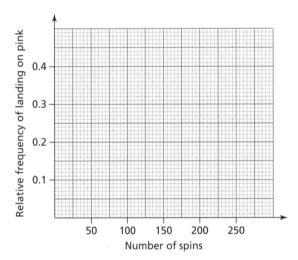

c) Write down the best estimate of the theoretical probability of the spinner landing on pink after 200 spins.

....

d) The spinner is spun a further 50 times and lands on pink 21 times.

   i) Find the relative frequency of the spinner landing on pink after 250 spins.

   ....

   ii) Plot this point on the graph.

   iii) What is the best estimate of the theoretical probability of the spinner landing on pink?

   ....

4 Jasmine suspects that a coin is biased. To find out she flips it 100 times and records the number of times it lands on a head after every 20 flips. Her results are shown in the table.

Number of flips	20	40	60	80	100
Number of heads	6	11	16	22	26
Relative frequency					

a) Complete the table for relative frequency.

b) Jasmine says the coin is biased.

   Is she correct? Give a reason for your answer.

   ....

   ....

c) Jasmine flips the coin in total 500 times.

   Find an estimate of the number of times it lands on *tails*.

   ....

5 Jim tests a coin to find out if it is biased.

He records the relative frequency of getting a head after tossing it a total of 20, 40, 60, 80 and 100 times.

He plots the relative frequencies on a graph, reproduced on the right.

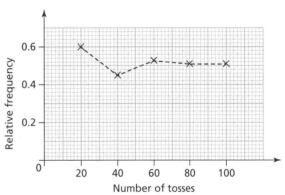

a) How many heads did Jim obtain in his first 20 tosses?

   ....

b) Do you think the coin is fair or biased? Explain your answer.

   ....

   ....

13 Equations and inequalities

You will practice how to:

- Understand that a situation can be represented either in words or as an equation. Move between the two representations and solve the equation (including those with an unknown in the denominator).
- Understand that a situation can be represented in words or as an inequality. Move between the two representations and solve linear inequalities.

13.1 Equations with fractions

Summary of key points

To solve an equation involving a fraction, you multiply both sides of the equation by the denominator. This then gives you an equation without a fraction, which you can solve.

For example: To solve

$$\frac{14}{x+3} = 2$$

Multiply both sides by $(x + 3)$ $\dfrac{14(x+3)}{(x+3)} = 2(x+3)$

Simplify $14 = 2(x + 3)$

Solve $7 = x + 3$

$x = 4$

Exercise 1

1   **Solve each of these equations to find the value of n.**

a) $\dfrac{n}{3} = 5$

b) $\dfrac{12}{n} = 4$

....

....

c) $\dfrac{42}{n} = 6$

d) $\dfrac{48}{3n} = 8$

....

....

2 Draw lines to match each equation with its correct solution.

| $\dfrac{3}{x+1}=1$ | $\dfrac{18}{x-1}=3$ | $\dfrac{16}{x-1}=-4$ | $\dfrac{21}{2x+1}=3$ | $\dfrac{35}{4x-11}=7$ |

| $x=-3$ | $x=4$ | $x=2$ | $x=7$ | $x=3$ |

3 Solve each of these equations to find the value of m.

a) $\dfrac{30}{2m+3}=6$

b) $\dfrac{18}{m+8}=3$

....

....

c) $\dfrac{100}{3m-7}=5$

d) $\dfrac{6}{3m+1}=3$

....

....

4 Solve each of these equations to find the value of t.

a) $\dfrac{15}{t}+6=3$

b) $\dfrac{30}{2t-3}+3=8$

> Make sure you have a single fraction on one side of your equation before multiplying by the denominator.

....

....

5 There are x children at a party. 5 more children join them.

There are 60 balloons at the party. The balloons are shared out equally among all the children.

Each child gets 5 balloons.

a) Write an equation in x to represent this situation.

....

b) Solve your equation to find the number of children originally at the party.

....

6 Draw a ring around the equation that is the odd one out.

$$\frac{8}{3x+5} = 4 \qquad \frac{6}{2-x} + 5 = 7 \qquad \frac{x}{2x+1} = 1 \qquad \frac{3}{5-x} - 4 = -3 \qquad \frac{x+5}{x} = -4$$

7 Lisa is trying to solve the equation $\dfrac{3p}{p-8} = 9$.

Here is her working. She has made a mistake.

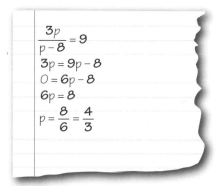

$$\frac{3p}{p-8} = 9$$
$$3p = 9p - 8$$
$$0 = 6p - 8$$
$$6p = 8$$
$$p = \frac{8}{6} = \frac{4}{3}$$

a) Draw a ring around her mistake.

b) Show the correct solution.

....

8 Sophie has $8n$ flowers. She wants to share the flowers among $n - 1$ friends.

She decides to include 3 more friends. With these friends included, each friend gets 9 flowers.

a) Write an equation in n to represent this situation.

....

b) Solve your equation to find how many friends Sophie was originally going to share the flowers with.

....

Think about

 9 Can a fraction equation have a fraction as a solution?

If you think the answer is yes, make up five fraction equations where the answer is a fraction.

13.2 Solving inequalities

Summary of key points

You can show inequalities on a number line.

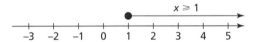 

You can solve inequalities in a similar way to equations.

Examples:

Solve: $4r - 3 < 29$

Solve: $-3t \geq 18$

$$4r - 3 < 29$$

$+3 \downarrow \qquad \downarrow +3$

$$4r < 32$$

$\div 4 \downarrow \qquad \downarrow \div 4$

$$r < 8$$

$$-3t \geq 18$$

$\div -3 \downarrow \qquad \downarrow \div -3$

$$t \leq -6$$

Note the change in the inequality sign after dividing through by -3. When you multiply or divide an inequality by a negative number, you need to reverse the inequality sign.

Exercise 2

1 Draw lines to match each inequality with its correct solution.

| $2x > 10$ | $x + 3 < 12$ | $x - 3 < 5$ | $\dfrac{x}{3} > 4$ | $3x < -15$ |

| $x < 8$ | $x < -5$ | $x > 5$ | $x < 9$ | $x > 12$ |

2 Solve these inequalities.

a) $5x < 35$

b) $x - 11 \geq -5$

....

....

c) $2x + 7 > 29$

d) $3(2x - 7) \leq 33$

....

....

3 Here are six inequalities. Solve each inequality and then write it in the correct column of the table.

The first inequality has been written in the table for you.

A
$5x - 2 > 3x + 2$

B
$7x - 1 < 4x + 8$

C
$-5x > -10$

D
$1 - 2x > -5$

E
$4 - 2x < x - 2$

F
$5 - x < 2$

Solution			
$x < 2$	$x > 2$	$x < 3$	$x > 3$
	A		

4 **a)** Solve the inequality $5n - 11 \leq n + 5$.

....

b) Show your answer to part **a** on the number line.

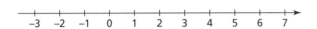

5 Draw lines to match each inequality with its correct number line diagram.

a) $-2 \leq n - 4 < 4$

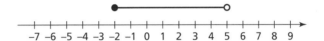

b) $1 \leq 2n + 3 < 15$

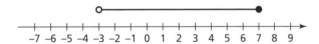

c) $0 < n + 3 \leq 10$

d) $-4 \leq 2n < 10$

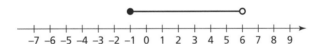

e) $-2 \leq \dfrac{n}{2} + 3 \leq 7$

6 Tina is growing tomato plants from seed. Her plants are 3.5 cm tall now. Tina expects the plants to grow 7 cm each week.

She will sell her tomato plants when they are more than 40 cm tall.

Tina writes this as an inequality. She uses w to represent the number of weeks.

a) Draw a ring around the correct inequality.

$7 + 3.5w > 40$ $7w + 3.5 \leq 40$ $7w + 3.5 > 40$ $7w + 3.5 \geq 40$ $7 + 3.5w \geq 40$

b) Explain how you know this is the correct inequality.

....

....

c) Solve the inequality and show your solution on the number line.

Solution ....

```
 +--+--+--+--+--+--+--+--+--+-->
 0  1  2  3  4  5  6  7  8  9
```

d) After how many weeks can Tina can sell her tomato plants?

....

7 Sonia has been solving inequalities. Put a tick (✓) in the box if Sonia's solution is correct and a cross (✗) if she has made a mistake.

If she has made a mistake, find the correct solution.

	Inequality	Sonia's solution		
a)	$10 - 3y < 25$	$y < -5$	☐	....
b)	$\frac{x}{2} + 3x \geq 14$	$x \geq 4$	☐	....
c)	$2x + 5 \leq 3 - 2x$	$x \leq -2$	☐	....
d)	$6 - \frac{1}{2}x \geq 2x + 1$	$x \geq 2$	☐	....
e)	$\frac{3}{4}x + 2 \leq \frac{1}{2} + 3x$	$x \leq \frac{2}{3}$	☐	....

Think about

8 Find five different inequalities that have the solution $x > 7$.

Calculations

14

You will practice how to:

- Use knowledge of the laws of arithmetic, inverse operations, equivalence and order of operations (brackets and indices) to simplify calculations containing decimals and fractions.

14.1 Simplifying calculations with decimals and fractions

Summary of key points

You can rewrite some calculations to make them easier to do, without changing the results. The table shows some methods you can use.

Method	Example
Converting between equivalent fractions and decimals	$0.68 \times 5^2 = \dfrac{68}{100} \times 25 = \dfrac{68}{4} = 17$
Changing the order of a calculation with additions and subtractions only, or multiplications and divisions only, when there are no brackets	$18 \times 17 \div 9 = 18 \div 9 \times 17 = 2 \times 17 = 34$
Noticing when there are **inverse operations** that cancel each other	$\sqrt{19^2} \times 7.1 \div 7.1 = 19 \times 7.1 \div 7.1 = 19$
Changing the order of numbers in addition or multiplication (addition and multiplication are **commutative**)	$4 \times \dfrac{5}{9} \times \dfrac{1}{2} = 4 \times \dfrac{1}{2} \times \dfrac{5}{9} = 2 \times \dfrac{5}{9} = \dfrac{10}{9}$
Changing the order of calculations by changing the grouping in addition or multiplication (addition and multiplication are **associative**)	$\left(0.79 \times \dfrac{1}{8}\right) \times 8 = 0.79 \times \left(\dfrac{1}{8} \times 8\right) = 0.79$
Using the **distributive law** to rewrite a multiplication by **partitioning** a number	$0.9 \times 46 = 1 \times 46 - 0.1 \times 46 = 46 - 4.6 = 41.4$
Using the distributive law in reverse to rewrite a multiplication	$\dfrac{5}{6} \times 8.2 + \dfrac{1}{6} \times 8.2 = \left(\dfrac{5}{6} + \dfrac{1}{6}\right) \times 8.2 = 8.2$

1 Use efficient methods to find:

a) $(1.7 + 5.49) + 8.3$

b) $\dfrac{3}{8} \times \dfrac{5}{7} \times \dfrac{4}{3}$

c) $\dfrac{3}{4} + 2.63 + \dfrac{5}{4}$

....

....

....

d) $(14.7 + 2.3) \times (7.7 + 2.3)$

e) $\left(\dfrac{5}{8} + 0.22\right) + \dfrac{3}{8}$

f) $1.25 \times \left(4 \times \dfrac{2}{9}\right)$

....

....

....

2 Here is Michael's receipt from the food store.

Use an efficient method to find the total cost of Michael's shopping.

$....................

FOOD STORE

Milk	$1.19
Juice	$1.64
Bread	$0.81
Apples	$1.25
Carrots	$0.36

3 Complete each calculation.

a) $......... + 9.2 - 4.7 = 9.2$

b) $12 \times \dfrac{11}{12} \div .... = 1$

c) $\dfrac{1}{8} \times \sqrt{(.........)^2} = 8$

d) $1.4 \div 9.3 \times 2 \times .... = 2.8$

e) $0.25 \times .... \times \dfrac{4}{5} = 2$

f) $1.75 + \dfrac{4}{5} + (0.2 - 1\dfrac{1}{4}) = ...$

4 Keith has done the calculations below.

$(0.5 + 0.3) \times 0.2 = 0.8 \times 0.2 = 0.16$

$\dfrac{2}{3} \times 3^2 = 2^2 = 4$

$10 \div \dfrac{2}{5} + \dfrac{3}{5} = 10 \div 1 = 10$

$\sqrt{\left(3.25 + 5\dfrac{3}{4}\right)} = \sqrt{\left(3\dfrac{1}{4} + 5\dfrac{3}{4}\right)} = \sqrt{9} = 3$

Mark Keith's work, and correct any mistakes.

5 Use efficient methods to find:

a) $1\frac{7}{10} + \frac{5}{8} + 2\frac{3}{10} - \frac{1}{8}$ b) $3.06 - 1.6 + 2.94 - 1.4$ c) $1.38 - 3.1 + 4.1 + 0.22$

....

d) $0.9 + 0.65 + \frac{1}{10} - \frac{3}{20}$ e) $2200 \times 0.34 \div 11$ f) $\frac{3}{4} \div \frac{7}{10} \times 4 \div 3$

....

6 Farzan wants to find $7.5 + \frac{1}{2} \times 1.6$.

His working is shown.

Describe the mistake Farzan has made.
Then find the correct answer.

....

....

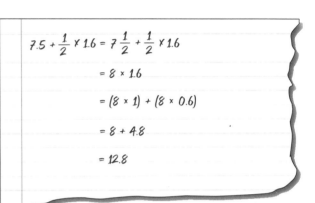

$$7.5 + \frac{1}{2} \times 1.6 = 7\frac{1}{2} + \frac{1}{2} \times 1.6$$
$$= 8 \times 1.6$$
$$= (8 \times 1) + (8 \times 0.6)$$
$$= 8 + 4.8$$
$$= 12.8$$

7 Use efficient methods to do these calculations.

a) $(1.6 \times 4.1) + (0.4 \times 4.1) = $ ....

b) $(1167 \times 0.04) - (0.04 \times 167) = $ ....

c) $3.1 \times 0.22 = $ ....

d) $5\frac{2}{3} \times 0.6 = $ ....

e) $\frac{3}{4} \times 8.2 - 0.25 \times 8.2$ ....

Think about

8 Describe when you can use the distributive law to make a calculation easier.

9 Complete this calculation so that:

a) the number in each box is a square number

$0.1 \times \boxed{} + 0.1 \times \boxed{} = 12.5$

b) one of the numbers is a negative number.

$0.1 \times \boxed{} + 0.1 \times \boxed{} = 12.5$

15 Pythagoras' theorem

You will practice how to:

- Know and use Pythagoras' theorem.

15. 1 Pythagoras' theorem

Summary of key points

Pythagoras' theorem states that the square of the hypotenuse in a right-angled triangle is equal to the sum of the squares of the other two sides.

This can be expressed as:

$a^2 + b^2 = c^2$

and, in terms of areas of squares, as:

area A + area B = area C

Pythagoras' theorem can be used to find the length of an unknown side in a right-angled triangle.

The side opposite the right angle in a right-angled triangle is called the **hypotenuse**.

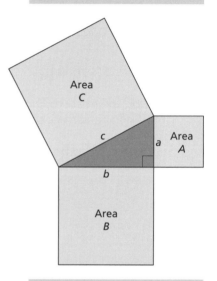

The diagrams in this exercise are not drawn to scale.

Exercise 1

1 Arrange these triangles into three groups.
Explain why you grouped the triangles the way you did.

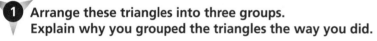

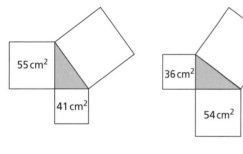

55 cm² 41 cm²

36 cm² 54 cm²

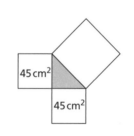

45 cm² 45 cm²

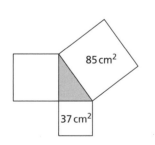

85 cm² 37 cm²

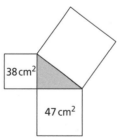

38 cm² 47 cm²

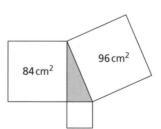

84 cm² 96 cm²

2 For each triangle, find the value of *x*.

a)

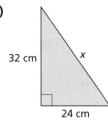

32 cm
x
24 cm

b)

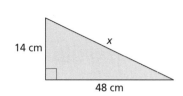

14 cm
x
48 cm

c)

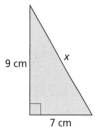

9 cm
x
7 cm

x = ……….. cm   *x* = ……….. cm   *x* = ……….. cm

3 Find the unknown side length in each triangle.

a)

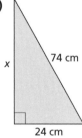

x
74 cm
24 cm

b)

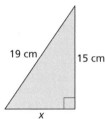

19 cm
15 cm
x

c)

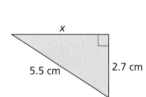

x
5.5 cm
2.7 cm

x = ……….. cm   *x* = ……….. cm   *x* = ……….. cm

4 Work out the length of each side marked with a letter.

a)

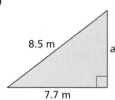

8.5 m
a
7.7 m

b)

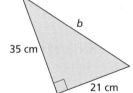

35 cm
b
21 cm

c)

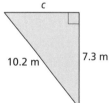

c
10.2 m
7.3 m

a = ……….. m   *b* = ……….. cm   *c* = ……….. m

5 Polly is trying to find the length of side *AB*.

Here is her working.

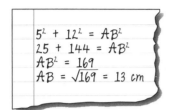

$5^2 + 12^2 = AB^2$
$25 + 144 = AB^2$
$AB^2 = 169$
$AB = \sqrt{169} = 13 \text{ cm}$

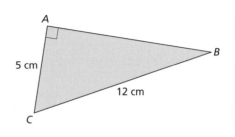

a) Polly looks at her answer and knows she has made a mistake. How can she tell this from her answer?

....

....

b) What mistake has Polly made in her working?

....

....

c) Correct Polly's working so that she finds the correct length of side *AB*.

....

....

6 Find the height of this isosceles triangle.

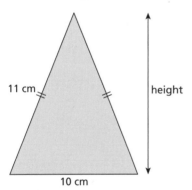

.... cm

7 Andy's house is 7 km due north of Bibi's house. Charlie's house is 4 km due east of Bibi's house. Find the distance between Andy's house and Charlie's house. Give your answer to 1 decimal place.

.... km

8 Calculate the perimeter of the trapezium. Give your answer to 1 decimal place.

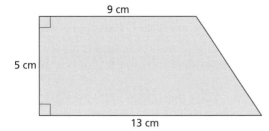

.... cm

9 Is triangle *PQR* a right-angled triangle? Explain your answer.

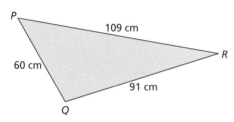

....

....

....

10 Calculate the length *x*.

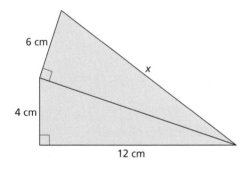

x = ... cm

Think about

11 Find three possible right-angled triangles with a hypotenuse of length 15 cm.

Can you find a right-angled triangle with integer side lengths with a hypotenuse of length 15 cm?

16 Measures of averages and spread

You will practice how to:

- Use mode, median, mean and range to compare two distributions, including grouped data.
- Interpret data, identifying patterns, trends and relationships, within and between data sets, to answer statistical questions. Make informal inferences and generalisations, identifying wrong or misleading information.

16.1 Median, mode and range for grouped distributions

Summary of key points

- The modal class is the class interval that has the highest frequency.
- The class containing the median is found by finding the position of the median value using $\frac{n+1}{2}$, where n is the number of data values.
- To find an estimate of the range from a grouped frequency table, subtract the smallest value from the first interval from the largest value in the last interval.

Exercise 1

1. Beth records the number of people queuing at a bus stop on 25 different occasions.

   a) Write down the modal class.

      ....

   b) Write down the class that contains the median value.

      ....

   c) Write down an estimate for the range.

      ....

Number of people in queue	Frequency
0–4	11
5–9	6
10–14	5
15–19	3
TOTAL	25

2. The table shows the monthly rainfall amounts in a city over the last 60 months.

   a) Write down an estimate for the range of the rainfall amounts.

      .... mm

Rainfall, x (mm)	Frequency
$20 \leq x < 40$	7
$40 \leq x < 60$	11
$60 \leq x < 80$	24
$80 \leq x < 100$	15
$100 \leq x < 120$	3

b) Write down the class interval that contains the median value.

....

3 The staff at a swimming pool record the ages of people using the pool.

The frequency diagrams show the ages of swimmers on one morning and one evening.

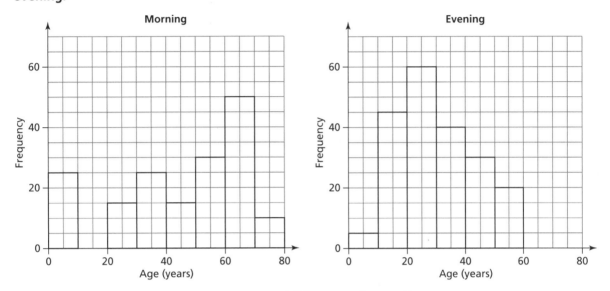

a) At what time of the day do more young children use the pool?

....

The median age of people using the pool in the morning is between 50 and 60 years.

b) Find the class interval that contains the median age of swimmers in the evening.

....

Jack concludes:

The swimmers in the evening are on average younger than those in the morning. The range of ages of the swimmers in the evening is greater than the range of the ages in the morning.

c) Comment on Jack's conclusion.

....

....

....

4 The frequency diagram shows the speeds of a sample of vehicles travelling on a road.

a) Write down the modal class.

....

b) Find how many vehicles were in the sample.

....

c) Write down the class interval that contains the median value.

....

d) Write down the maximum possible range of speeds.

....

Speeds of vehicles

5 Ahmed measures the handspan (in cm) of all the students in his class. Here are his results.

16.8	15.3	17.8	21.9	15.5	13.5	15.5	20.6	20.6	21.3
21.9	14.2	14.9	14.9	12.7	17.4	17.7	19.4	13.7	18.8
19.5	16.9	18.0	15.3	19.8	13.8	15.6	17.8	20.8	18.8

a) Complete the frequency table.

b) Write down the class interval that contains the median value.

....

c) Write down the modal class.

....

Handspan, s (cm)	Frequency
$12 \leq s < 15$	7

d) Compare the maximum range from the table with the actual range.

....

....

6 The frequency polygon shows the times taken for a class of students to run 100 m.

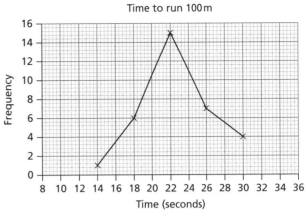

Time to run 100 m

a) How many students are in the class?

....

b) Find the modal class interval.

....

c) Find the class interval that contains the median.

....

d) Find the maximum range of times.

....

Think about

7 To find the class intervals from the frequency polygon draw a frequency diagram on top of it with the crosses at the midpoint of each bar.

How would you change a frequency diagram into a frequency polygon?

16.2 Mean from a grouped frequency distribution

Summary of key points

- When data has been grouped, the exact data values are not known so only an **estimate of the mean** can be found.
- To estimate the mean in a grouped frequency table, find the midpoint of each class interval then multiply the frequencies by the midpoints.

> Add columns for the midpoints of the class intervals and 'Midpoint × frequency', and find the totals.

Example: Find an estimate of the mean length.

Length, l (cm)	Frequency	Midpoint	Midpoint × frequency
$5 \le l < 10$	2	7.5	15
$10 \le l < 15$	10	12.5	125
$15 \le l < 20$	16	17.5	280
$20 \le l < 25$	6	22.5	135
$25 \le l < 30$	3	27.5	82.5
TOTALS	37		637.5

Estimate of mean $= \dfrac{\text{sum of (midpoint} \times \text{ frequency)}}{\text{sum of frequencies}}$

$\qquad\qquad\quad = \dfrac{637.5}{37}$

$\qquad\qquad\quad = 17.2 \, \text{cm (1 d.p.)}$

Exercise 2

1 Complete the frequency table and use it to find an estimate of the mean weight.

Weight, w (kg)	Frequency	Midpoint	Midpoint × frequency
$2 \le w < 6$	2		
$6 \le w < 10$	10		
$10 \le w < 14$	16		
$14 \le w < 18$	6		
$18 \le w < 22$	3		
TOTALS			

.... kg

2 The frequency table shows the ages of 80 people.

Age, a (years)	Frequency		
$15 < a \leq 20$	6		
$20 < a \leq 25$	18		
$25 < a \leq 30$	26		
$30 < a \leq 35$	12		
$35 < a \leq 40$	11		
$40 < a \leq 45$	5		
$45 < a \leq 50$	2		

Find an estimate of the mean age.

.... years

3 The frequency diagram shows the heights of a class of students.

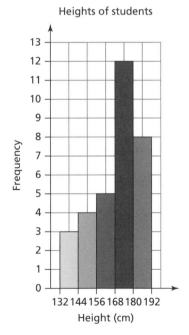

Heights of students

a) Find how many students were in the class.

....

b) Write down the modal class.

....

c) Find an estimate of the mean height of students in the class.

.... cm

④ **The frequency table shows the durations of 56 telephone calls made by a call centre worker.**

Time, t (minutes)	Frequency
$0 < t \le 2$	5
	18
	22
$6 < t \le 8$	
$8 < t \le 10$	2

a) Complete the frequency table.

b) Write down the modal class.

....

c) Find an estimate of the mean call duration.

.... minutes

d) Explain why the mean is an estimate.

....

....

5 The frequency polygon shows the time spent exercising each week by 50 gym members.

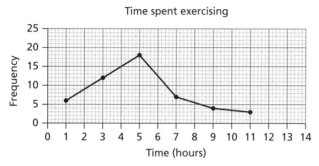

a) Complete the frequency table.

Time, t (hours)	Frequency
$0 < t \leq 2$	6
$2 < t \leq 4$	

b) Find an estimate of the mean time spent exercising.

.... hours

6 The frequency diagram shows the weights in kilograms of
50 visitors to a health clinic.

a) The clinic manager says, 'The median weight of visitors to
the clinic is 90 kg.'

Comment on the manager's claim.

....

....

b) Calculate an estimate of the mean weight of visitors to the clinic.

Weights of visitors
to clinic

.... kg

c) Explain why the mean weight is an estimate.

....

....

Percentages

You will practice how to:

- Understand compound percentages.

17.1 Compound percentages

Summary of key points

There are two ways to answer **compound percentage** questions.

Example:

The value of a car is $16 000 and this decreases by 20% each year.
Calculate the value of the car after 3 years.

Method 1
Value after 1 year = $16 000 × 0.80 = $12 800
Value after 2 years = $12 800 × 0.80 = $10 240
Value after 3 years = $10 240 × 0.80 = $8192

Method 2
Value after 3 years = $16 000 × 0.80^3 = $8192

Exercise 1

1 Draw lines to match each percentage change with the correct multiplier.

Percentage change	Multiplier
0% increase	0
10% decrease	0.1
10% increase	0.9
90% decrease	1
100% decrease	1.1
100% increase	2

2 a) The price of a book in a shop is $8.00. The shop increases the price by 10% and then decreases the new price by 25%.

Find the final price of the book.

$..........................

b) In another shop, the price of a book is $8.00. The shop decreases the price by 25% and then increases the new price by 10%.

Find the final price of the book.

$..........................

Think about

3 Compare the two answers in question 2.

Is it always true that the order of the percentage changes does not affect the final value?

4 A mobile phone costs $50. Its price then increases by 20%. Later, the new price is reduced by 20%.

Draw a ring around the final price.

$32 $48 $50 $72

5 To calculate each percentage change below, what should you multiply the original quantity by?

The first answer has been written for you.

Percentage change	Multiply by
5% increase per day, for 3 days	1.05^3
80% decrease per hour, for 7 hours	
100% increase per year, for 4 years	
1% decrease per month, for 12 months	

6 The price of a new model of mobile phone is $200. The price decreases by 15% each month.

Find the price of the phone after 9 months. Round your answer to the nearest dollar.

$..........................

7 Since the year 2010, the population of a city has been increasing by 2% every year.

The population in 2010 was 620000.

Find the population in 2020. Round your answer to the nearest thousand.

....

8 The population of an endangered species of animal is 960. It is predicted to decrease by 12% each year.

Dilys says, 'If the prediction is correct, the population will be less than 300 after 8 years.'

Is she correct? Show your working.

....

....

9 The value of a motorcycle decreases by 10% each year. Eivind wants to calculate the value after five years.

He says, '5 × 10% = 50%, so the value halves in five years.'

Is Eivind correct? Explain your answer.

....

....

18 Sequences

You will practice how to:

- Generate linear and quadratic sequences from numerical patterns and from a given term-to-term rule (any indices).
- Understand and describe nth term rules algebraically (in the form $an \pm b$, where a and b are positive or negative integers or fractions, and in the form $\frac{n}{a}$, n^2, n^3 or $n^2 \pm a$, where a is a whole number).

18.1 Generating sequences

Summary of key points

You can generate a sequence by identifying the term-to-term rule.

In a **linear sequence** the terms increase or decrease by the same amount each time.

In a **quadratic sequence** the **second difference** is the same each time.

Example:

0.5,   1,   2.5,   5,   8.5,

+0.5 +1.5 +2.5 +3.5

+1   +1   +1

The term-to-term rule is 'add 0.5, add 1.5, add 2.5, add 3.5'.

The second difference is +1 each time.

This is a quadratic sequence.

The 6th term is 8.5 + 4.5 = 13.

The 7th term is 13 + 5.5 = 18.5.

Exercise 1

1 Find the next two terms in each sequence.

a) −4, −2, 0, 2, 4, 6, ..., ...

b) 0, −4, −8, −12, −16, ..., ...

c) 3, 13, 24, 36, 49, ..., ...

d) −1, 4, 14, 29, ..., ...

e) 3, 6, 11, 18, ..., ...

f) −3, 0, 5, 12, 21, ..., ...

2 Complete each sequence to match the term-to-term rule.

   a) Square and add 2   2   ....

   b) Add 3, add 4, add 5   12   ....

   c) Multiply by $\frac{2}{5}$ and add 2   402   ....

   d) Square and add 1   5   26   ....

3 Draw a line to match each sequence with its term-to-term rule.

Multiply by 2 and subtract 1	3, 4, 11, 116, ...
Square and subtract 6	3, 3, 3, 3, ...
Square and subtract 8	−2, −4, 8, 56, ...
Square and subtract 5	−2, −5, −11, −23, ...

4 Draw a ring around the odd one out.

   A 6, 14, 24, 36, ...   **B** −1, 0, 3, 8, 15, ...   **C** −1, 5, 11, 17, ...   **D** 2, 6, 12, 20, ...

   Explain your answer.

   ....

   ....

   ....

5 Find the missing terms in these linear sequences.

   a) 21, 19, ... , ... , 13

   b) 6, ... , ... , ... , 30, ...

   c) .... , 5, ... , ... , ... , ... , 50

   d) .... , 25, ... , ... , ... , ... , 5, ...

6 Ashley says that the sequence 1, 4, 7, 10, 13, ... is quadratic.

   Is she correct?   Yes ☐   No ☐

   Explain why.

   ....

   ....

   ....

7 **a)** Write in the missing number:

The term-to-term rule of the sequence 1, 3, 11, …is 'square and add …….. '.

b) What is the position in the sequence of the first number greater than 1000?

....

> **Think about**
>
> **8** 3, 8, 15, 24, 35, … is a quadratic sequence.
>
> Explain why.

18.2 nth term rules

Summary of key points

The **nth term rule** of a **linear sequence** has the form $an \pm b$, where a and b are positive or negative integers or fractions.

The nth term rule of a **quadratic sequence** contains n^2 and no higher powers.

The nth term rule of a **cubic sequence** contains n^3 and no higher powers.

You can generate sequences by using nth term rules of linear, quadratic or cubic sequences.

For example, the 10th term of the sequence with nth term rule $n^2 + 3$ is $10^2 + 3 = 103$.

In general, you can identify a sequence with an nth term rule of the form $n^2 + a$ by comparing the terms with the square number sequence, n^2.

For example, in the sequence 2, 5, 10, 17, 26, … each term is one more than its corresponding term in the square number sequence. So the nth term rule for this sequence is $n^2 + 1$.

n^2	1	4	9	16	25
$n^2 + 1$	2	5	10	17	26

Exercise 2

1 By comparing each sequence with the square number sequence, n^2, write the nth term rule for each sequence.

a) 6, 9, 14, 21, … nth term = ………….….

b) 10, 13, 18, 25, … nth term = ………….….

c) −2, 1, 6, 13, … nth term = ………….….

d) −6, −3, 2, 9, … nth term = ………….….

2 Draw a line to match each sequence with its *n*th term rule.

−1, 2, 7, 14, ...	$n^2 + 4$
−5, −2, 3, 10, ...	$n^2 + 8$
5, 8, 13, 20, ...	$n^2 − 6$
9, 12, 17, 24, ...	$n^2 − 2$

3 Find the 4th term in the sequence given by each *n*th term rule.

a) $3n + 7$ ....

b) n^3 ....

c) n^2 ....

d) $n^2 − 2$ ....

e) $n^2 + 5$ ....

4 Find the 5th term in the sequence generated by each *n*th term rule.

a) $11 − 0.5n$ ....

b) $n^2 + 4$ ....

c) $n^2 − 5$ ....

d) n^3 ....

5 a) Two terms of the cubic sequence with *n*th term rule n^3 add up to 280.

What are the two terms?

.... and ....

b) Two terms of the sequence with *n*th term rule $\frac{n}{4}$ add up to 1.

What are the two terms?

.... and ....

6 Write the numbers 8, 11, 40, 64, 77, 85 and 125 in the correct positions in this table.

	In sequence with *n*th term $n^2 + 4$	Not in sequence with *n*th term $n^2 + 4$
In sequence with *n*th term n^3		
Not in sequence with *n*th term n^3		

7 Terms in Sequence A have nth term rule $120 - 5n$.

Terms in Sequence B have nth term rule n^3.

Ben says that there is no term in Sequence A that is equal to the 5th term in Sequence B.

Is he correct?   Yes ☐   No ☐

Explain your reasoning.

....

....

....

8 a) Find the nth term rule for each sequence.

Sequence A   −4, −1, 4, 11, ...

nth term rule = ...

Sequence B   −4, −1, 2, 5, 8,...

nth term rule = ...

b) Which term in Sequence B is equal to the 7th term in Sequence A?

....

9 a) Draw the next pattern.

b) How many dots are there in pattern n?

....

19 Area and measures

You will practice how to:

- Know and use the formulae for the area and circumference of a circle.
- Estimate and calculate areas of compound 2D shapes made from rectangles, triangles and circles.
- Know and recognise very small or very large units of length, capacity and mass.

19.1 Area and circumference of a circle

Summary of key points

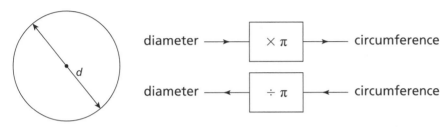

Circumference = πd
Circumference = $2\pi r$

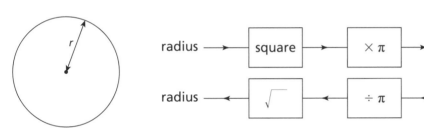

Area = πr^2

Exercise 1

1 Calculate the area and the circumference of this circle. Give your answers to 3 significant figures.

17 cm

Area = ... cm²

Circumference = ... cm

2 For her homework Lola-belle calculated the area of some circles. Mark her homework using ticks (✓) and crosses (✗).

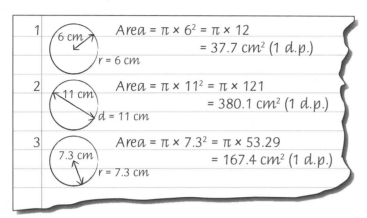

1 6 cm $Area = \pi \times 6^2 = \pi \times 12$
   r = 6 cm $= 37.7\ cm^2\ (1\ d.p.)$

2 11 cm $Area = \pi \times 11^2 = \pi \times 121$
   d = 11 cm $= 380.1\ cm^2\ (1\ d.p.)$

3 7.3 cm $Area = \pi \times 7.3^2 = \pi \times 53.29$
   r = 7.3 cm $= 167.4\ cm^2\ (1\ d.p.)$

3 Complete the table.

Radius	Diameter	Area (rounded to nearest whole number)
4 cm		
7.8 cm		
	17 cm	

4 Mia has a train set. The track is a circle with diameter 1.4 metres.

Mia's train goes round the track 25 times.

Calculate the total distance the train travels.

Give your answer to the nearest metre.

.... m

5 The area of a circle is 115 cm².

Calculate the radius of the circle. Give your answer to 2 decimal places.

.... cm

6 The diagram shows a running track. The 'ends' of the track are semicircles.

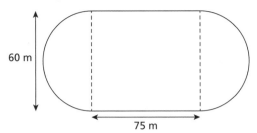

Phillipe wants to run at least 4 km and must run a whole number of laps.

What is the least number of laps he can run?

....

7 Shapes A, B, C and D are shown in the diagram.

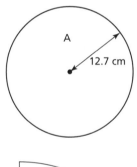

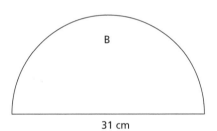

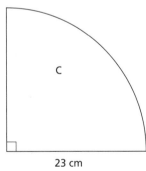

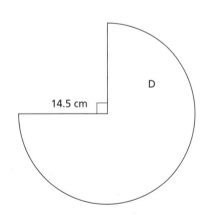

Write the shapes A to D in the correct position in this Carroll diagram.

	Perimeter less than 80 cm	Perimeter not less than 80 cm
Area less than 450 cm^2		
Area not less than 450 cm^2		

8 Shape A is made from semicircles and Shape B is made from semicircles and straight lines.

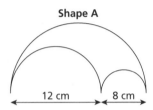

Shape A

12 cm   8 cm

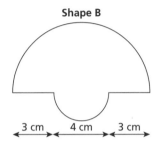

Shape B

3 cm   4 cm   3 cm

a) Find the perimeter of each shape.

Perimeter of A = ... cm   Perimeter of B = ... cm

b) Show that Shape A has a larger area than Shape B.

....

....

....

....

....

....

Think about

9 Jez thinks that the circumference of a circle of diameter 2n cm is the same as the total of the circumferences of n circles each of diameter 2 cm.

He also thinks that the area of a circle of radius n cm is the same as the total of the areas of n circles each of radius 1 cm.

By considering different values of n, explore whether Jez is correct.

Summary of key points

A compound shape is a shape made up of two or more different shapes.

The area of a compound shape is found by calculating each area separately and combining areas.

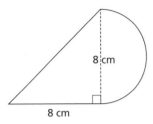

8 cm

8 cm

Area of the above compound shape = area of triangle + area of semicircle

Exercise 2

1 **Calculate the area of this shape. Give your answer to the nearest whole number.**

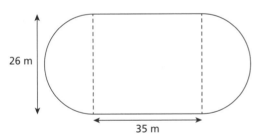

26 m

35 m

.... m²

2 Calculate the area of this shape. Give your answer to 1 decimal place.

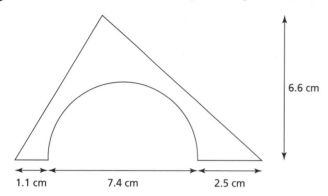

6.6 cm

1.1 cm   7.4 cm   2.5 cm

.... cm²

3 The diagram shows a semicircle drawn inside a rectangle.

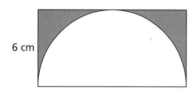

6 cm

Work out the size of the shaded area.

.... cm²

4 Tom paints a rim around the edge of a circular plate, as shown in the diagram below.

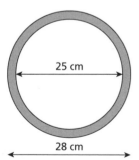

25 cm

28 cm

Calculate the shaded area that he paints.

.... cm²

5 A shape is formed by cutting a semicircle from a square.

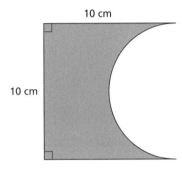

10 cm

10 cm

Calculate the area, giving your answer correct to 1 decimal place.

.... cm²

6 Show that nearly 75% of the rectangle is shaded.

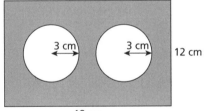

3 cm   3 cm   12 cm

18 cm

....

....

....

....

....

....

7 The diagram shows a clock face, made up of a semicircle on top of a trapezium.

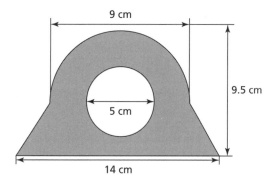

Calculate the area of the shaded part. Give your answer to 1 decimal place.

.... cm²

19.3 Small and large units

Summary of key points

1 light year = 9 460 730 472 580 800 m   1 tonne (t) = 1000 kg

Small units:

milli (m) ... is 1 thousandth (10^{-3})

micro (μ) ... is 1 millionth (10^{-6})

nano (n) ... is 1 billionth (10^{-9})

Large units:

mega (M) ... is 1 million (10^{6})

giga (G) ... is 1000 million =
1 billion (10^{9})

tera (T) ... is 1 million million (10^{12})

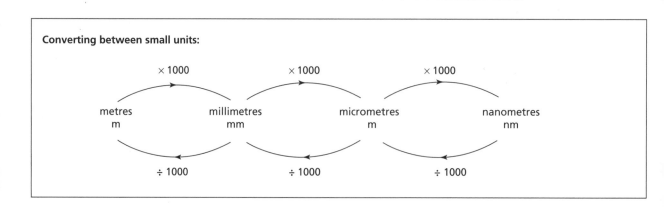

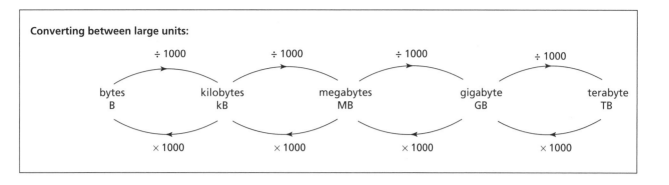

Converting between large units:

bytes B $\xrightarrow{\div 1000}$ kilobytes kB $\xrightarrow{\div 1000}$ megabytes MB $\xrightarrow{\div 1000}$ gigabyte GB $\xrightarrow{\div 1000}$ terabyte TB

$\times 1000$ (reverse direction)

Exercise 3

1 Draw a ring around the larger measurement in each set.

a) 1 tonne   956 kg

b) 1 light year 9×10^{16} metres

c) 8 GB   8200 MB

d) 1 microgram   1 nanogram

2 Draw a ring around all the measurements that are equal to 720 t.

72 000 kg   720 000 kg   0.072 kt   0.00072 Mt   0.000072 Gt

3 Tick to show if these statements are true or false.

	True	False
0.75 megatonne = 0.75 million tonnes	☐	☐
73 000 nm = 7.3 μm	☐	☐
430 GB = 0.43 TB	☐	☐
125 mg = 0.000125 g	☐	☐
280 000 B = 28 MB	☐	☐

4 Draw a ring around the most appropriate unit.

a) Radius of the Milky Way   nanometre   metre   light year

b) Mass of an elephant   gram   tonne   megatonne

c) Volume of a teardrop   litre   microlitre   gigalitre

d) Storage capacity of a laptop hard drive   gigabyte   byte   kilobyte

5 Write each set of measurements in order of size, starting with the smallest.

a) 0.08 kg   80 µg   80 t   800 mg   800 g

....
smallest   largest

b) 75 mm   0.75 cm   75 µm   0.0075 km   750 nm

....
smallest   largest

6 Complete these conversions.

a) 5400 GB = ... TB   **b)** 290 kg = ... tonne

c) 0.45 µm = ... nm   **d)** 0.029 g = ... µg

e) 3 GB = 3000.....................   **f)** 0.00012 mm = 120 ....

7 A grain of sand has a mass of 4.4 mg.

How many grains of sand would you expect in a sandcastle with a mass of 2 kg?

Give your answer to 2 significant figures.

....

8 In 2018, carbon dioxide emissions from burning fossil fuels were 37 gigatonnes.

a) Write 37 gigatonnes in tonnes.

.... tonnes

b) Write your answer in standard form.

.... tonnes

9 Amber transfers a video file that is 28.5 GB from a memory card to a computer hard drive.

It takes 1 second to download 145 MB.

Find how long it takes to transfer the whole video file. Give your answer to the nearest minute.

.... minutes

Presenting and interpreting data 2

You will practice how to:

- Record, organise and represent categorical, discrete and continuous data. Choose and explain which representation to use in a given situation:
  - o dual and compound bar charts
  - o pie charts
  - o scatter graphs
  - o infographics.
- Interpret data, identifying patterns, trends and relationships, within and between data sets, to answer statistical questions. Make informal inferences and generalisations, identifying wrong or misleading information.

20.1 Pie charts and bar charts

Summary of key points

Information shown in one statistical diagram can sometimes be presented in a different form.

For example, the compound bar chart shows information about the numbers of people in cars at 9 a.m. and 11 a.m. The information for 9 a.m. has been converted into a pie chart.

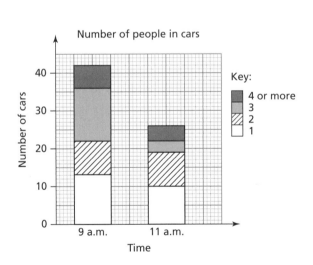

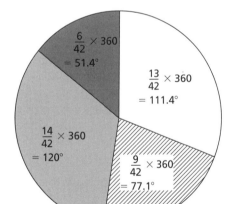

Number of people in cars at 9 a.m.

1 **Real data question** The compound bar graph shows information about the highest qualifications of people aged 25–29 in Singapore in different years.

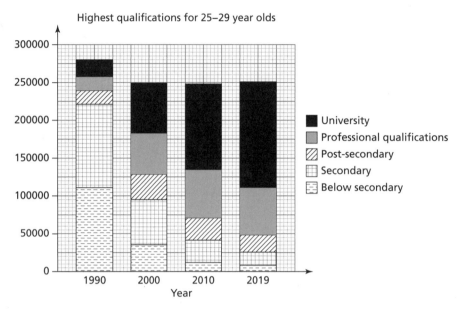

Source: Contains information from Education, Language Spoken and Literacy, accessed on 15/09/20 from https://www.singstat.gov.sg/ which is made available under the terms of the Singapore Open Data Licence version 1.0 {https://data.gov.sg/open-data-licence}

a) Write down the percentage of 25–29-year-olds whose highest qualification in **2019** was University.

....

b) Find the percentage of 25–29-year-olds whose highest qualification in **2000** was Secondary.

....

c) Make **two** comments about how the highest qualifications have changed over time.

(1) ....

....

(2) ....

....

2 Pedro runs exercise classes four days each week. He has classes in the morning and in the afternoon.

The dual bar chart shows the numbers of people attending his classes one week. One bar is missing.

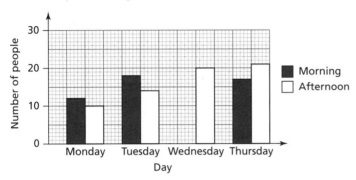

People attending Pedro's exercise classes

a) Pedro had 26 people in his exercise class on Wednesday morning. Complete the dual bar chart.

b) Find how many more people attended classes in the morning than in the afternoon.

....

c) Redraw the graph as a compound bar chart.

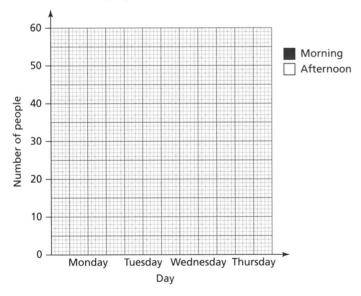

3 The bar chart shows the most popular social media platforms for a sample of students.

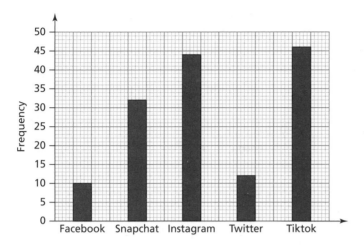

a) Find the number of students in the sample.

....

b) Draw a pie chart to represent the data.

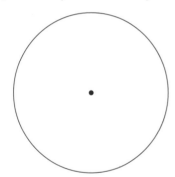

4 The table shows sales (in $) of coffee, tea and hot chocolate at a café during the five days it is open each week.

	Coffee	Tea	Hot chocolate
Saturday	38	24	24
Monday	27	8	25
Tuesday	7	34	8
Wednesday	32	38	11
Thursday	15	28	12

Caitlin draws this compound bar chart to show the information.

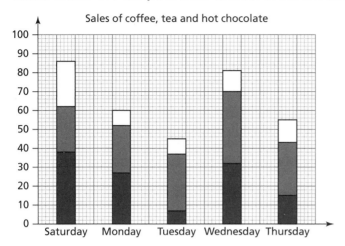

Write down three errors Caitlin has made when drawing her graph.

(1) ....

....

(2) ....

....

(3) ....

....

5 Kabir shared $420 with friends Raj and Ananya in the ratio of the number of letters in their names.

a) Find how much each of the three friends received.

Kabir $ ....   Raj $ ....   Ananya $ ....

b) Draw a pie chart to represent this information.

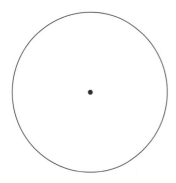

Summary of key points

Infographics can show a range of statistical information. They are frequently used to show data (such as population data) about countries.

For example, the chart here shows the numbers of males and females in different age groups.

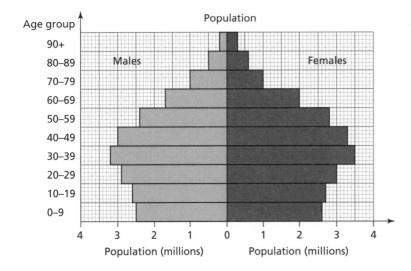

1 The diagram shows some information about the ages of people living in a town. Two bars are missing.

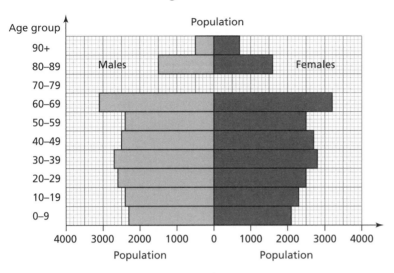

a) There are 2300 males and 2700 females in the town aged 70–79 years. Complete the diagram.

b) Calculate the number of females aged between 40 and 59 years.

....

c) Find how many more males than females are aged under 10 years.

....

2 The diagram shows how the number of mobile phones (per 100 people) has changed in three countries between 1990 and 2018.

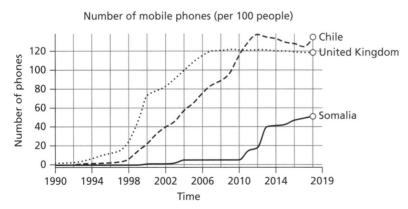

Source: The World Bank: Mobile cellular subscriptions (per 100 people): International Telecommunication Union (ITU) World Telecommunication/ICT Indicators Database.

a) Write down which of the countries had the most mobile phones (per 100 people) in 2014.

....

b) Write down the year when Somalia had approximately the same number of mobile phones (per 100 people) as Chile did in 2000.

....

c) Write down the year when mobile phone ownership (per 100 people) was approximately the same in Chile and in the United Kingdom.

....

d) The population of Somalia in 2018 was approximately 15 million.

Estimate the number of mobile phones in Somalia in 2018. Show how you worked out your answer.

....

e) Describe the pattern in mobile phone ownership in the United Kingdom between 1990 and 2018.

....

....

....

3 The infographic shows information about the population of Singapore in 2019 and how it has changed since 2010.

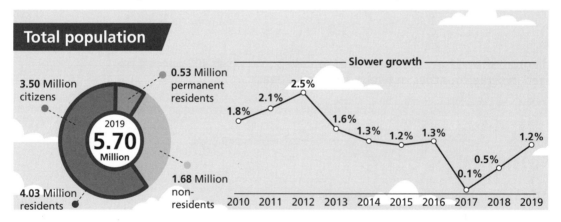

Source: Contains information from Population and Population Structure, accessed on 15/09/20 from https://www.singstat.gov.sg/ which is made available under the terms of the Singapore Open Data Licence version 1.0 [https://data. gov.sg/open-data-licence]

a) Write down the number of residents in Singapore in 2019.

....

b) Find the percentage of the total population who are citizens.

.... %

c) Describe the growth of population between 2017 and 2019 and compare it with the growth of population between 2010 and 2012.

....

....

....

20.3 Correlation

Summary of key points

Two sets of data show **correlation** if there is a relationship between them.
Correlation can be positive or negative.

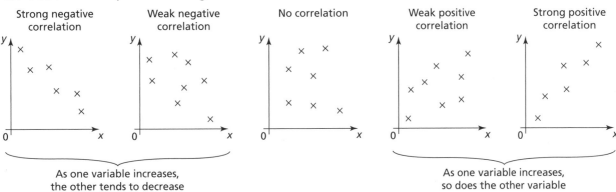

Strong negative correlation | Weak negative correlation | No correlation | Weak positive correlation | Strong positive correlation

As one variable increases, the other tends to decrease

As one variable increases, so does the other variable

Interpolation uses a line of best fit to estimate a value that is within the data points.
Extrapolation uses a line of best fit to estimate a value outside the plotted data points.

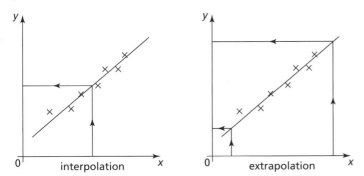

interpolation

extrapolation

We cannot assume that the data will always follow this pattern, so **extrapolation** can be unreliable.

Causation is when a change in one variable **causes** a change in the other.

1 Draw a ring around the diagrams that show positive correlation.

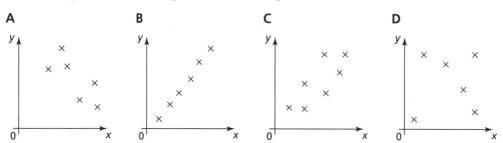

A   B   C   D

2 What type of correlation would you expect between these pairs of variables?

	Positive correlation	Negative correlation	No correlation
Temperature and coat sales	☐	☐	☐
Shoe size of 14-year-olds and time they spend sleeping	☐	☐	☐
Number of rooms in a house and house price	☐	☐	☐

3 Pria surveys the students in her school.

She produces these three scatter diagrams to show her results.

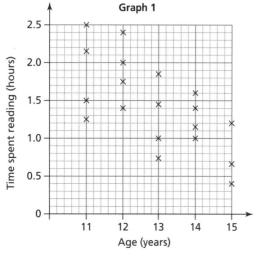

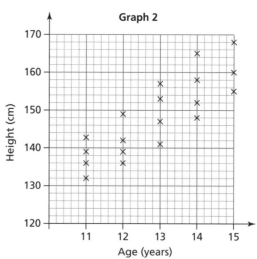

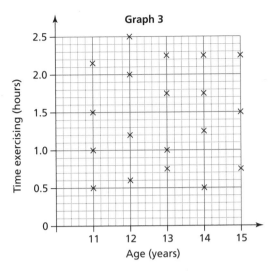

Graph 3

a) Which of Pria's graphs shows no correlation?

....

b) What does Graph 1 show about the relationship between the time spent reading and age?

....

c) What type of correlation is shown in Graph 2?

....

4 A teacher collects information about the number of minutes her students spent reading last week and the mark they scored in a reading test.

Her results are shown on the scatter graph.

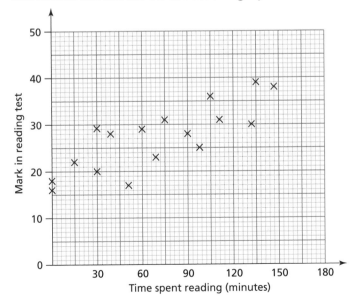

a) Describe the correlation between the mark in the reading test and the time spent reading.

....

....

b) Draw a line of best fit on the scatter graph.

c) Estimate the mark in the reading test for a student who spent:

i) 120 minutes reading ....   **ii)** 3 hours reading ....

d) Which of your answers to part **c** is likely to be more reliable? Give a reason for your answer.

....

....

5 Gavin has been asked to explore the following question:

Is there a link between a person's height and their arm span?

He collects data from a sample of 10 people.

Height (cm)	160	154	161	173	156	168	177	164	158	170
Arm span (cm)	153	147	154	165	157	154	179	164	163	168

a) Draw a suitable diagram to show these data.

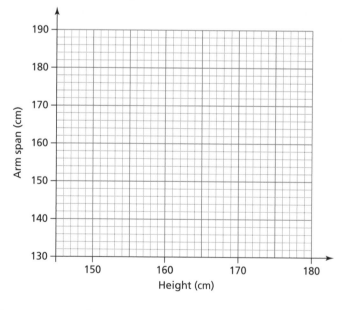

b) Write a conclusion related to the question Gavin was asked to explore.

....

....

6 The graph shows the number of doctors and broadband speed for 10 countries

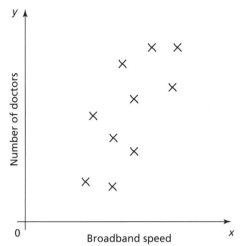

a) Describe the correlation shown in the graph.

   ....

b) Decide whether or not an increase in the broadband speed will cause a change in the number of doctors.

   ....

Think about

7 Look at the labels on some food products in your kitchen cupboards or fridge at home. Investigate whether food products that contain more fat also tend to contain more sugar. Draw a scatter graph to show your results.

21 Ratio and proportion

You will practice how to:

- Use knowledge of ratios and equivalence for a range of contexts.
- Understand the relationship between two quantities when they are in direct or inverse proportion.

21.1 Using ratios in context

Summary of key points

You can use ratios to solve problems, such as the two examples below.

Example:	Example:
Samia and Joao share some chocolates in the ratio 3 : 5. Joao gets 8 more chocolates than Samia. Find the total number of chocolates.	Franz and Jung share some chocolates in the ratio 2 : 3. Jung gets 12 chocolates. Find the number of chocolates Franz gets.
Joao gets 2 more shares than Samia, since 5 – 3 = 2. 2 shares equal 8 chocolates. So 1 share equals 4 chocolates. The total number of shares is 3 + 5 = 8. So the total number of chocolates is 8 × 4 = 32.	Jung gets 3 shares. 3 shares equal 12 chocolates. So 1 share equals 4 chocolates. Franz gets 2 shares. So Franz gets 2 × 4 = 8 chocolates.

Exercise 1

 1 **A bag contains red, blue, green and orange balls in the ratio red : blue : green : orange = 1 : 3 : 4 : 6.**

Tick the correct boxes to show whether each statement is true, false, or impossible to say from the information given.

	True	False	Impossible to say
The least common colour of ball is red.	☐	☐	☐
There are 5 more orange balls than red balls.	☐	☐	☐
There are twice as many orange balls as blue balls.	☐	☐	☐

2 Brad and Chanya share some apples in the ratio 3 : 5. Chanya gets 4 more apples than Brad gets.

Find the number of apples Brad gets.

....

3 Two numbers are in the ratio 2 : 3. One of the numbers is 12.

Find the two possible values of the other number.

.... or ....

4 Grace and Hiroe each have some sweets. Their numbers of sweets are in the ratio 20 : 13.

'If Grace gives Hiroe 2 sweets, the ratio of their sweets will be 6 : 5.'

Is this statement definitely true, possibly true, or definitely false? Explain your answer.

....

....

....

5 Lembit, Morten and Nisha share some money in the ratio 3 : 2 : 2. Lembit gets $4.50.

Find the total amount of money shared.

$ ....

6 Jiajun cuts a piece of wood into four lengths, in the ratio 6 : 5 : 3 : 2. The longest piece is 20 cm longer than the shortest piece.

Find the length of the longest piece.

.... cm

7 A bag contains red, yellow and green sweets, in these ratios:

red : yellow = 3 : 2   and   yellow : green = 1 : 5

Find the ratio of red : green sweets.

red : green = ………….. : …………..

Think about

8 Amelia is 15 years old and Bahram is 6 years old.

Write two different ratio questions about Amelia's and Bahram's ages. The questions should not give both of their ages.

21.2 Direct and inverse proportion

Summary of key points

If two **variables** are **directly proportional**, their values are always in the same ratio. If the value of one variable is multiplied by a number, the value of the other variable is multiplied by the same number.

If two variables are **inversely proportional**, the product of their values is always the same. If the value of one variable is multiplied by a number, the value of the other variable is divided by the same number.

Exercise 2  Q1–5,7

1 In an experiment, Harper measures values of the mass and volume of some pieces of copper. The mass and volume are directly proportional.

a) She has made a mistake in one row of the table. Draw a ring around this row.

b) Explain how you know that the mistake is in this row.

Mass (g)	Volume (cm³)
18	2
27	3
45	5
90	9

………………………………………………………………………………………………………

………………………………………………………………………………………………………

2 In a different experiment, Leroy measures the density and volume of some blocks of equal mass, made of different metals. In this experiment the density and volume are inversely proportional because the blocks are of equal mass.

Density (g/cm³)	Volume (cm³)
2	16
3	12
4	9
9	4

a) He has made a mistake in one row of the table. Draw a ring around this row.

b) Explain how you know that the mistake is in this row.

....

....

3 The area of a rectangle is fixed, but the length and width are variable.

Are the length and width directly proportional, inversely proportional, or neither? Explain your answer.

....

....

4 8 British pounds = 84 Hong Kong dollars.

Change 1680 Hong Kong dollars to British pounds.

.... British pounds

5 It takes 3 tractors 20 hours to plough some fields. How long would it take:

a) 15 tractors to plough the same fields?

.... hours

b) 5 tractors to plough the same fields?

.... hours

6 Six concert tickets for adults cost $171. Four concert tickets for children cost $69.

Find the cost of concert tickets for five adults and three children.

$ ....

7 Tick the correct boxes to show which statements are true if two variables *x* and *y* are directly proportional, and which are true if *x* and *y* are inversely proportional.

> If you are not sure, try experimenting with possible values of *x* and *y*.

	True if *x* is directly proportional to *y*	True if *x* is inversely proportional to *y*
If *x* is multiplied by a number, *y* is divided by the same number.	☐	☐
If *x* is divided by a number, *y* is divided by the same number.	☐	☐
x is always the same proportion (fraction) of the total.	☐	☐
y is always the same proportion (fraction) of the total.	☐	☐
The product of *x* and *y* always has the same value.	☐	☐
The fraction $\frac{x}{y}$ always has the same value.	☐	☐

22 Relationships and graphs

You will practice how to:

- Use knowledge of coordinate pairs to construct tables of values and plot the graphs of linear functions, including where y is given implicitly in terms of x ($ax + by = c$), and quadratic functions of the form $y = x^2 \pm a$.
- Understand that straight-line graphs can be represented by equations. Find the equation in the form $y = mx + c$ or where y is given implicitly in terms of x (fractional, positive and negative gradients).
- Understand that a situation can be represented either in words or as a linear function in two variables (of the form $y = mx + c$ or $ax + by = c$), and move between the two representations.
- Read, draw and interpret graphs and use compound measures to compare graphs.

22.1 Drawing linear and quadratic graphs

Summary of key points

If the equation of a **linear function** has the form $y = mx + c$, then m represents the gradient and c is the y-intercept.

Some linear functions are not written with y as the subject. Instead, they may be written **implicitly**, for example $3x + 2y = 12$. You can rearrange these functions in the form $y = mx + c$ to identify the gradient and the y-intercept.

$y = x^2$ and $y = x^2 + 3$ are examples of simple **quadratic functions**.

To draw the graph of a quadratic function, you need to begin by drawing a table of values.

For example, here is a table of values and graph of $y = x^2$.

x	−3	−2	−1	0	1	2	3
y	9	4	1	0	1	4	9

plot points
and join with a smooth curve

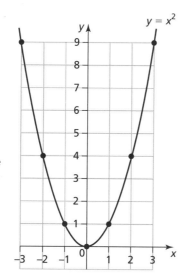

The graph of a quadratic function has the shape of a **parabola**.

1 Here are the equations of two straight lines:
$2y = 3x - 4$ and $y + 2x = 8$.

a) Make y the subject of each equation.

.... and ....

b) Complete the table of values for the two lines.

$2y = 3x - 4$

x	−2	0	2	4
y				

$y + 2x = 8$

x	−2	0	2	4
y				

c) Draw the two straight lines on the grid.

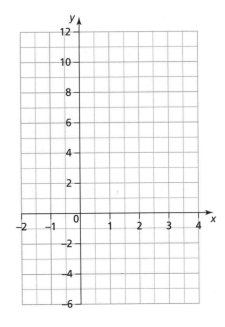

2 $x + 3y = 6$ and $5x + 2y = 10$ are both equations of straight lines.

a) Complete the tables of values.

$x + 3y = 6$

x	0	3	6
y			

$5x + 2y = 10$

x	0	1	2
y			

b) Draw the two straight lines on the grid.

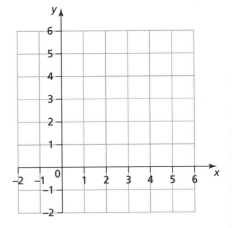

3 **a)** Find where the graph of $4x + 3y = 24$ intercepts the x- and y-axes.

   x-intercept ....

   y-intercept ....

b) Use your answers to part **a** to draw the graph of $4x + 3y = 24$.

c) Draw the graph of $2x + 7y = 14$ on the same axes.

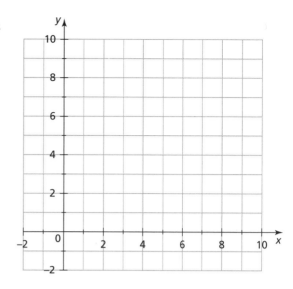

4 **Match each linear function to its graph.**

$x - 4y = 4$ ....

$4x + y = 8$ ....

$6x + 5y = 30$ ....

$3y - 2x = 9$ ....

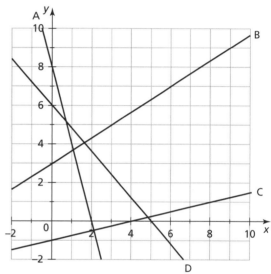

5 **a)** Complete the table of values for $y = x^2 - 1$.

x	-3	-2	-1	0	1	2	3
y	8						

b) Draw the graph of $y = x^2 - 1$.

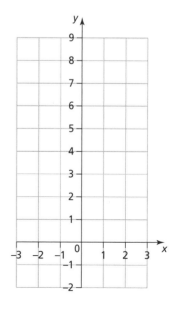

6 a) Use six numbers from this list to complete the table.

8	5	6	10	13	5	8	13	6

x	−3	−2	−1	0	1	2	3
$y = x^2 + 4$				4			

b) Draw the graph of $y = x^2 + 4$.

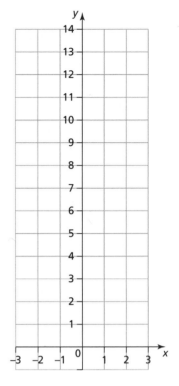

7 Zainah says that the graph drawn on this grid has equation $3x + 2y = 9$.

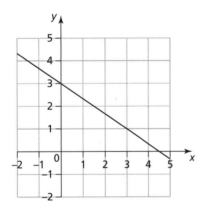

Explain how you can tell she is wrong.

....

....

Summary of key points

Finding the equation of a line

Example: Find the equation of the line.

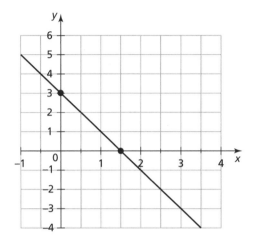

y-intercept = 3

Gradient = $\frac{-3}{1.5}$ = −2

Equation of line is $y = -2x + 3$

Finding the gradient and y-intercept of a line

Example: Find the gradient and y-intercept of the line $2x - 3y = 15$.

Rearrange to make y the subject.

$$2x - 3y = 15$$

+3y $\downarrow$ $\qquad$ $\downarrow$ +3y

$$2x = 3y + 15$$

−15 $\downarrow$ $\qquad$ $\downarrow$ −15

$$2x - 15 = 3y$$

÷3 $\downarrow$ $\qquad$ $\downarrow$ ÷3

$$y = \frac{2x}{3} - 5 = y$$

So $y = \frac{2x}{3} - 5$

Therefore gradient is $\frac{2}{3}$ and y-intercept is −5.

Exercise 2

1 Draw a ring around the lines with gradient 2 and y-intercept 5.

$2x + y = 5$ $\qquad$ $2x - y + 5 = 0$ $\qquad$ $y - 5x = 2$ $\qquad$ $2y - 2 = 4x + 8$

2 Draw a ring around the line that is the odd one out.

$y = \frac{1}{4}x + 2$ $\qquad$ $4y = x - 8$ $\qquad$ $4y + x = 2$ $\qquad$ $2x - 8y = 24$

Give a reason for your choice.

....

3 Tick the statement about the graph of $26y + 78x = 65$ that is true.

It has gradient -3 and y-intercept 2.5 ☐

It has gradient $-\frac{1}{3}$ and y-intercept 2.5 ☐

It has gradient $-\frac{1}{3}$ and y-intercept -2.5 ☐

It has gradient 3 and y-intercept -2.5 ☐

4 Write down the equation of each graph.

A ....

B ....

C ....

D ....

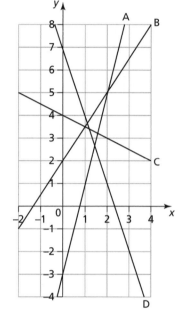

5 Myla draws the line $y = 2x - 1$.

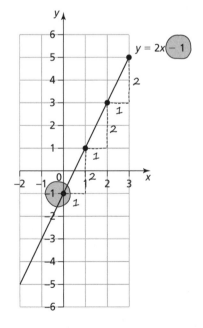

$y = 2x - 1$

a) Explain Myla's method.

....

....

....

b) Use Myla's method to draw these two lines on the axes below.

    i) $y = 4x - 2$

    ii) $y = 5 - 3x$

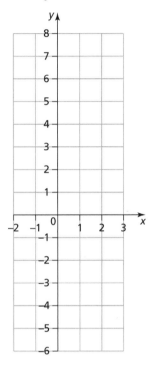

6 Here are the equations of two linear functions:

$3y - x = 2$ and $2y - 5x = 1$

Find which function has the steeper graph. Show your working.

....

Summary of key points

Some real-life situations can be described in words or expressed as a linear function in two variables.

Example: Small notebooks cost $3 each and large notebooks cost $5 each. A customer spends $500 buying x small notebooks and y large notebooks.

A linear function describing this situation is $3x + 5y = 500$.

Exercise 3

1. A company produces black pens and blue pens. They produce exactly 500 pens each day.

   x is the number of black pens produced in a day and y is the number of blue pens produced in a day.

   Draw a ring around the linear function that describes this situation.

   $y = 500 + x$ $x + y = 500$ $x - y = 500$ $xy = 500$

2. Vanessa uses strips of wood to make bird boxes and storage boxes. A bird box needs 1 metre of wood. A storage box needs 2 metres of wood.

   The total length of wood she has is 150 metres. She wants to use all of her wood.

   Write a linear function that can describe this situation. Define any variables you introduce.

   ....

   .... represents ....

   .... represents ....

3. A party organiser needs exactly 200 fireworks for a party. Fireworks are sold in small or big boxes.
   There are 6 fireworks in a small box. There are 8 fireworks in a big box.

   Write a linear function to represent this situation. Define the variables you use.

   ....

   .... represents ....

   .... represents ....

4 Susie makes and then paints picnic tables and benches.

It takes Susie 5 hours to make a picnic table. It takes 4 hours to make a bench.

She wants to spend exactly 100 hours on making items.

Let *x* represent the number of picnic tables and *y* represent the number of benches made.

a) Write a linear function to represent this situation.

....

It takes 2 hours for Susie to paint a picnic table and 1 hour to paint a bench. She wants to spend exactly 34 hours on painting items.

b) Write a linear function to represent this situation.

....

c) Jo says that Susie could produce 20 picnic tables. Explain why this cannot be true.

....

....

Think about

5 ...*x* + ...*y* = 180 is a linear function, where *x* is the number of adults and *y* is the number of children.

Write two different situations where this linear function can be used.

Give the linear functions for each situation.

22.4 Real-life graphs

Summary of key points

Exchange rates can be used to convert between two different currencies. A **conversion graph** can be drawn to help do the conversion.

The gradient of a real-life graph can represent a **rate of change**.

The gradient of a distance–time graph gives the **speed** of an object. Speed is measured in units like metres per second or kilometres per hour.

A **compound measure** is a quantity that is formed from two different measures. Speed is an example of a compound measure.

1 The graph shows the cost (C dollars) of ordering *n* books from Company A.

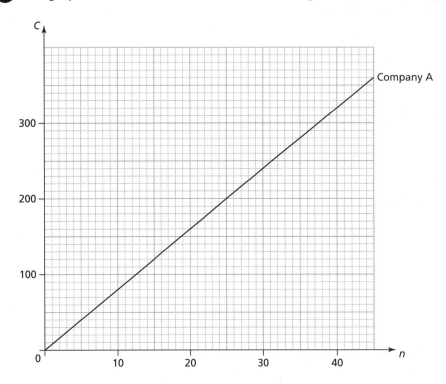

a) Calculate the gradient of the line. Explain what it represents in this context.

Gradient ....

....

b) The cost of ordering books from Company B is given by the formula
   $C = 6.5n$.

   Draw a graph for Company B on the axes above.

c) Explain whether it is cheaper to order books from Company A or Company B.

   Give a reason for your answer.

   ....

   ....

2 Stav leaves a city at 09 00. Alina leaves the same city 30 minutes later.

They both drive a distance of 240 km to an airport.

Their journeys are shown on the graph below.

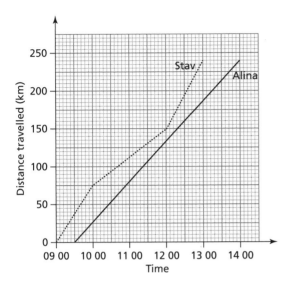

a) Calculate the average speed in km/h for Stav's journey.

.... km/h

b) Whose average speed for the journey was greater? ....

Show how you made your decision.

....

....

3 Simone uses one of two different routes to travel to work each day.

Last week, she used one route on Monday and the other route on Tuesday. Her journeys are shown in the following graphs.

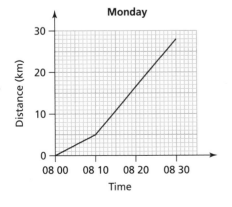

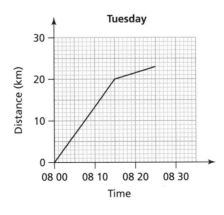

a) Work out Simone's average speed for her journey to work on Monday.

.... km/min

b) On which day was Simone's average speed faster?

....

Show your working.

....

....

....

4 Toni fills his inflatable pool with water.

The graph shows the depth of water in the pool plotted against time.

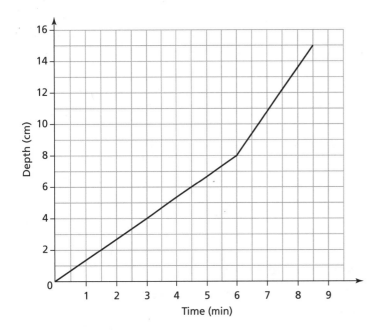

a) Calculate the rate at which the depth of water increases during the first 6 minutes.

Give units in your answer.

....

b) Toni says, 'The rate of increase in the depth of water for the final 2.5 minutes is greater than for the first 6 minutes.'

Is he correct?   Yes ☐   No ☐

Give a reason for your answer.

....

....

5 The graphs show conversions between pounds (£) and US dollars ($) and between euros (€) and US dollars.

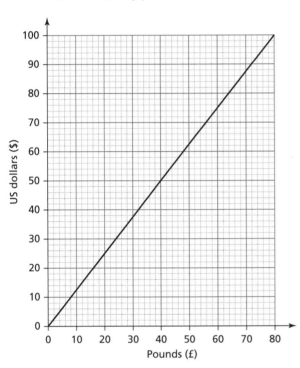

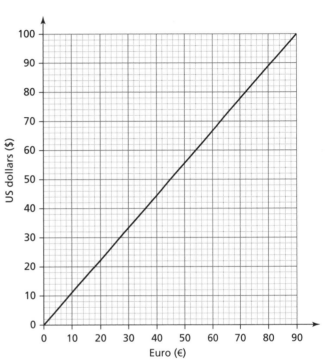

a) Complete this conversion rate between £ and $.

£1 = $ ....

b) Ashley changes £96 into US dollars.

She spends half of her dollars on a meal.

She converts the money she has left into euros.

Find how many euros Ashley receives.

€ ....

6 **The graph shows Tia's journey to a shop and back.**

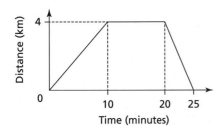

a) Calculate the speed Tia travelled to the shop.
   Give your answer in km/minute.

.... km/min

b) Calculate Tia's speed on her way home from the shop.
   Give your answer in km/hour.

.... km/hour

23 Probability 2

You will practice how to:

- Identify when successive and combined events are independent and when they are not.
- Understand how to find the theoretical probabilities of combined events.

23.1 Independent events

Summary of key points

Two events are called **independent** if they do not affect each other. The probability of one of the events does not depend on whether the other happens or not.

Events that are not independent are called **dependent**.

Exercise 1

1 Tamara is a train driver.

P is the probability that the train she drives tomorrow will arrive at its final station on time.

Q is the probability that the train she drives tomorrow leaves its starting station on time.

R is the probability that she has eggs for breakfast tomorrow.

a) Write down a pair of events that are likely to be independent. Give a reason for your answer.

.... and ....

because ....

b) Write down a pair of events that are likely to be dependent. Give a reason for your answer.

.... and ....

because ....

2 Lewis has a 5-sided spinner numbered 1, 2, 3, 4 and 5. He spins the spinner and throws a coin.

Here are some events.

A	B	C	D
He gets an even number on the spinner.	He gets a Head on the coin.	He gets a 5 on the spinner.	He gets a Tail on the coin.

Tick to show if each pair of events are independent or dependent.

	Independent events	Dependent events
A and B	☐	☐
A and C	☐	☐
B and D	☐	☐
C and D	☐	☐

3 Simona throws an ordinary dice twice.

X is getting a 6 on the first throw.

Y is getting a 6 on the second throw.

Are X and Y independent?　　Yes ☐　　No ☐

Explain your answer.

....

....

4 Raj has 8 tins in his cupboard. 5 of the tins contain fruit.

He takes a tin out of the cupboard at random and eats what is inside. He then takes a second tin out of the cupboard at random.

F is getting a tin containing fruit on the first pick.
G is getting a tin containing fruit on the second pick.

Raj says, 'F and G are independent events as I am picking at random.'

Give a reason why Raj is not correct.

....

....

23.2 Tree diagrams and calculating probabilities

Summary of key points

Multiplication rule: P(A and B) = P(A) × P(B) provided that A and B are independent.

Probabilities for combined events are sometimes found from **tree diagrams**.

- Write the probabilities for each event on the branches. The probabilities on each set of branches should add to 1.
- If the events are independent, the probabilities for combined events can be found by multiplying along the branches.

1 The probability that Kev has eggs for breakfast is 0.2.

The probability that he listens to the radio at breakfast is 0.7.

Assuming independence, find the probability that at breakfast
he eats eggs and listens to the radio.   ....

2 Jim takes a counter at random from a bag containing 10 counters, numbered
1, 2, ..., 10.

Lyn throws an ordinary dice.

Find the probability that

a) Jim and Lyn both get a 3.   ....

b) Jim gets an even number and Lyn gets an odd number.   ....

c) Jim gets a number less than 4 and Lyn gets a prime number. ....

3 Molly spins these two spinners.

She adds together the numbers her spinners land on.

a) Complete the sample space diagram to show the possible total scores.

5-sided spinner

4-sided spinner		2	2	3	3	3
	3	5	5	6		
	4	6				
	4					
	4					

b) Use the sample space diagram to find the probability that her total score is **7**.

....

c) Complete the tree diagram to show the outcomes and probabilities.

5-sided spinner	4-sided spinner	Combined event	Probability

$\frac{1}{4}$ 3   2 and 3 $\frac{2}{5} \times \frac{1}{4} = \frac{2}{20}$ or $\frac{1}{10}$

2

$\frac{2}{5}$

....   4   2 and 4   ....

3   3 and 3   ....

....

....   3

....   4   3 and 4   ....

d) Use your tree diagram to find the probability that the spinners show the same number.   ....

4   **Dave and Gina each independently have either an apple or a banana for their lunch.**

The probability that Dave has an apple is 0.85.

The probability that Gina has an apple is 0.4.

a) Complete the tree diagram by writing a probability on each branch.

Dave   Gina

apple

....

apple

....

....

banana

apple

....

....

banana

....

banana

b) Find the probability that both Dave and Gina have a banana.

....

c) Find the probability that they have different fruits.

....

5 Abed has two bags containing coloured counters.

Bag 1  Bag 2

Key:
G = green
Y = yellow
R = red

He takes a counter at random from each bag.

a) Complete the tree diagram to show the probabilities.

Bag 1   Bag 2

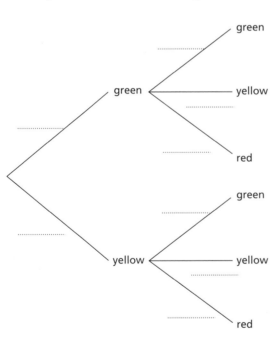

green
....
green
yellow
....
....
red

green
....
yellow
yellow
....
....
red

....

....

....

b) Find the probability that he gets a green counter on both picks.

....

c) Find the probability that at least one of the counters he picks is green.

....

6 Meera spins each of these spinners once.

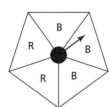 

Key:
R = red
B = blue
G = green
Y = yellow

Draw lines to match the outcomes to the probability calculations.

| Red and Green | $\frac{2}{5} \times \frac{3}{7}$ |

| Red and Yellow | $\frac{3}{5} \times \frac{4}{7}$ |

| Blue and Green | $\frac{2}{5} \times \frac{4}{7}$ |

| Blue and Yellow | $\frac{3}{5} \times \frac{3}{7}$ |

7 An 8-sided dice has 6 blue faces and 2 yellow faces. The dice is thrown twice.

a) Draw a tree diagram to show the possible outcomes from the throws.

Label each branch with its probability.

b) Find the probability of getting two yellow faces.

....

c) Find the probability of getting at least one blue face from the two throws.

....

Think about

8 Make up a probability question to match this calculation.

$$\frac{5}{8} \times \frac{4}{9}$$

3D shapes

You will practice how to:

- Use knowledge of area and volume to derive the formula for the volume of prisms and cylinders. Use the formula to calculate the volume of prisms and cylinders.
- Use knowledge of area, and properties of cubes, cuboids, triangular prisms, pyramids and cylinders to calculate their surface area.
- Identify reflective symmetry in 3D shapes.

24.1 Volumes of prisms

> Remember that the same shape (called the **cross-section**) runs through the entire length of a **prism**.
>
> A **cylinder** has a circular cross-section.

Summary of key points

Volume of prism = area of cross-section × length
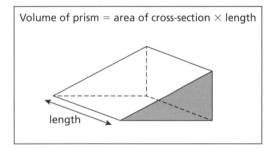

Volume of cylinder = area of circular base × height
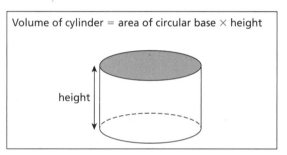

Exercise 1

> The diagrams in this exercise are not drawn to scale.

1 Use the cross-sectional area to calculate the volume of each prism.

a)

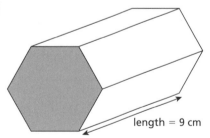

length = 9 cm

cross-sectional area = 70 cm²

.... cm³

b)

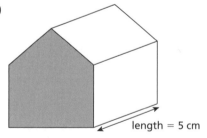

length = 5 cm

cross-sectional area = 64 cm²

.... cm³

2 Choose the correct volume for each cylinder. Answers should be given to 2 significant figures.

| 250 cm³ | 1600 cm³ | 2000 cm³ | 2200 cm³ |

| 2400 cm³ | 2600 cm³ | 2900 cm³ | 10 000 cm³ |

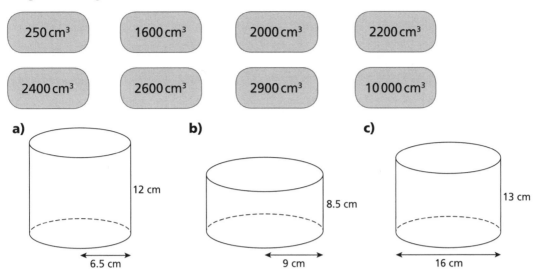

a)

12 cm

6.5 cm

b)

8.5 cm

9 cm

c)

13 cm

16 cm

.... cm³   .... cm³   .... cm³

3 Find the difference between the volumes of these prisms.

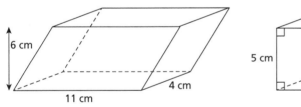

6 cm

11 cm

4 cm

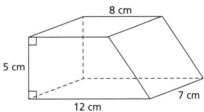

8 cm

5 cm

12 cm

7 cm

.... cm³

4 Calculate the unknown dimensions.

a)

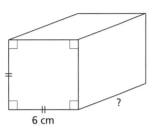

6 cm

Volume = 270 cm³

Length = ... cm

b)

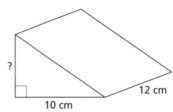

?   10 cm   12 cm

Volume = 480 cm³

Height = ... cm

c)

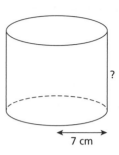

?   7 cm

Volume = 2000 cm³

Height = ... cm

5 These two prisms are equal in volume. Find the value of *a*.

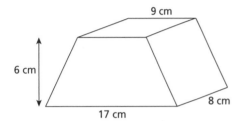

9 cm

6 cm

17 cm

8 cm

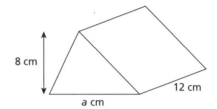

8 cm

a cm

12 cm

a = ...

6 A cylinder has a volume of 360 cm³ and a height of 11 cm.
Padmini says that the radius is 3 cm to the nearest cm.

Is Padmini correct?   Yes ☐   No ☐

Show how you worked out your answer.

....

....

....

....

....

Think about

7 Design a prism that has a volume equal to 400 cm³.

The cross-section of your prism should be a trapezium.

Summary of key points

The surface area of a prism can be found by adding together the area of each face.

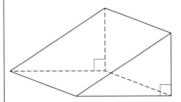

Here, three faces are rectangles and two faces are triangles.

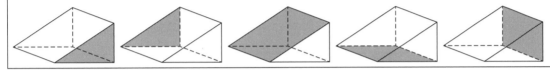

The surface area of a cylinder can be found by adding the area of the top and bottom to the area of the curved surface.

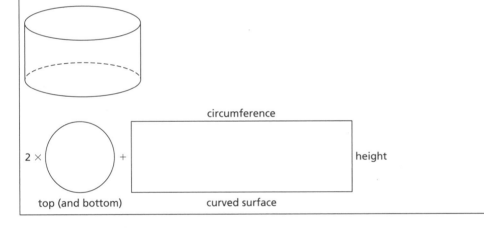

$2 \times$ ◯ $+$ ▭

top (and bottom)   circumference   curved surface   height

Exercise 2

1 **Calculate the surface area of this triangular prism.**

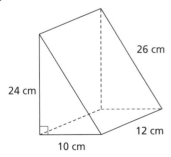

26 cm

24 cm

12 cm

10 cm

.... cm²

2 Here are four prisms.

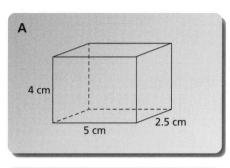

A

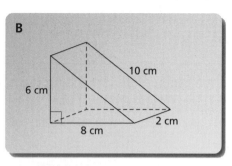

B

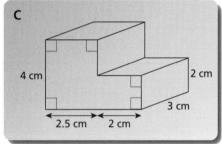

C

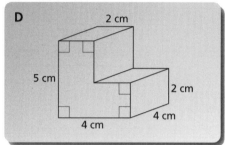

D

Match each prism to the correct surface area.

79 cm²   85 cm²   96 cm²   100 cm²

A = ...   B = ...   C = ...   D = ...

3 The diagram shows a prism with a trapezium as cross-section.

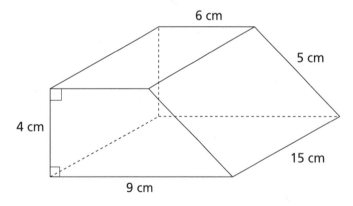

Calculate the surface area of this prism.

.... cm²

4 Here is a triangular prism.

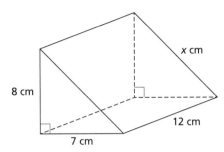

a) Use Pythagoras' theorem to calculate the value of *x*. Give your answer to 2 decimal places.

....

b) Calculate the surface area of this prism. Give your answer to 1 decimal place.

.... cm²

5 Here are two cylinders.

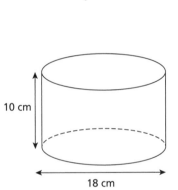

Which of these cylinders has the larger surface area? A ☐   B ☐

Show how you worked out your answer.

....
....
....
....
....

6 The diagram shows a pyramid with a rectangular base.

The vertex of the pyramid is vertically above the centre of the rectangle.

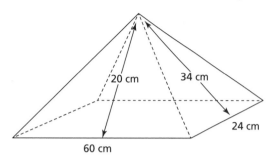

Show that the surface area of the pyramid is 3456 cm².

....

....

....

....

....

7 The total surface area of this prism is 236 cm².

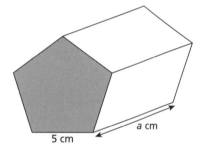

The cross-section of the prism is a regular pentagon with an area of 43.0 cm² (to 1 decimal place).

Calculate the value of a.

$a = $ ....

Summary of key points

A **plane of symmetry** divides a 3D shape into halves.
The two halves are congruent and are mirror images of each other.

For example, a cuboid has three planes of symmetry.

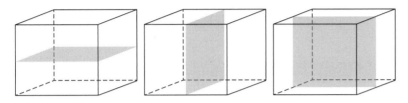

Exercise 3

1 **Tick (✓) the shapes if the shaded section shows a plane of symmetry.**

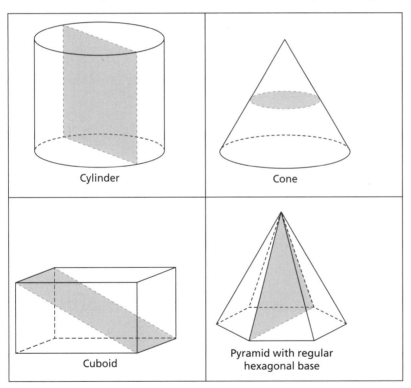

2 Write down the number of planes of symmetry in each 3D shape.

a)

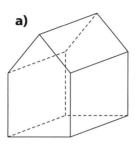

....

b)

....

3 Sketch the four planes of symmetry for a square-based pyramid.

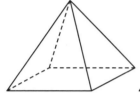  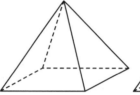 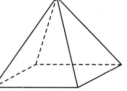

4 The diagram shows part of a shape and its plane of symmetry. Draw the whole shape on the grid.

5 Sketch three different prisms with a triangular cross-section with the stated property.

a) One plane of symmetry

b) Two planes of symmetry

c) Four planes of symmetry

6 Show how four cubes can be joined together to make a 3D shape with five planes of symmetry.

25 Simultaneous equations

You will practice how to:

- Understand that the solution of simultaneous linear equations:
  - o is the pair of values that satisfy both equations
  - o can be found algebraically (eliminating one variable)
  - o can be found graphically (point of intersection).

25.1 Solving simultaneous equations with like coefficients

Summary of key points

Simultaneous equations can be solved by eliminating one variable from the equations.

Example:

Solve: $3x + 2y = 11$ ①

$x + 2y = 5$ ②

$$
\begin{array}{rcl}
3x + 2y &=& 11 \\
- \quad x + 2y &=& 5 \\
\hline
2x \quad\quad &=& 6
\end{array}
$$

Both equations contain $2y$, so subtract them to eliminate y.

$2x = 6$ ① − ②

$x = 3$

Substitute $x = 3$ into equation ②.

$3 + 2y = 5$

So $2y = 2$

$y = 1$

> **Simultaneous equations** are two (or more) equations containing the same unknowns that are solved at the same time so the solutions satisfy all the equations.

Exercise 1

1 In each part, substitute the values of x and y into the equations and decide whether or not they are a correct solution.

			Correct solution	Incorrect solution
a)	$x = 3, y = 2$	$x + y = 5$ $x - y = 1$	☐	☐
b)	$x = 2, y = 7$	$2x + y = 11$ $y - x = 6$	☐	☐
c)	$x = 5, y = 1$	$x + 3y = 8$ $2x - y = 9$	☐	☐
d)	$x = 0, y = 3$	$x + 2y = 6$ $4x - y = -3$	☐	☐

2 Solve each pair of simultaneous equations using subtraction.

a) $3x + y = 9$
$x + y = 5$

b) $3x + 5y = 49$
$3x + 2y = 25$

$x = ...$

$y = ...$

$x = ...$

$y = ...$

3 Solve each pair of simultaneous equations using addition.

a) $3x + 2y = 17$
$5x - 2y = 7$

b) $3x + 5y = 57$
$-3x + y = -3$

$x = ...$

$y = ...$

$x = ...$

$y = ...$

4 Solve by elimination:

a) $x + y = 5$
$x + 3y = 9$

b) $2x + 3y = 10$
$2x + 5y = 18$

$x = ...$

$y = ...$

$x = ...$

$y = ...$

c) $3x - y = 16$
    $5x - y = 30$

d) $2x - 3y = 19$
    $x + 3y = -4$

$x = \ldots\ldots\ldots\ldots$

$y = \ldots\ldots\ldots\ldots$

$x = \ldots\ldots\ldots\ldots$

$y = \ldots\ldots\ldots\ldots$

5 **Here is Khalish's attempt to solve**

$4m - n = 6$

$2m + n = 0$

simultaneously.

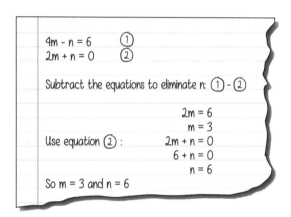

a) Draw a ring around the **first** line of working which is incorrect.

b) Work out the correct solutions of the simultaneous equations.

$m = \ldots\ldots\ldots\ldots$

$n = \ldots\ldots\ldots\ldots$

6 When you solve simultaneous equation such as

$x = 20 - y$

$x - y = 4,$

What would you do first? Why?

25.2 Solving simultaneous equations with unlike coefficients

Summary of key points

When the variables in a pair of simultaneous equations have different coefficients, you will need to multiply one or both of the equations by a constant before you can eliminate one of the variables.

Example:

Solve: $3x + 4y = 23$ ①

       $x - 2y = 1$ ②

Multiply equation ② by 3 so both equations contain $3x$.

$x - 2y = 1$ ② $\xrightarrow{\times 3}$ $3x - 6y = 3$ ③

Subtract equations ① and ③ to eliminate x.

$3x + 4y = 23$ ①

$3x - 6y = 3$ ③

   $10y = 20$ ① − ③

     $y = 2$

Substitute $y = 2$ into equation ①.

$3x + 8 = 23$

So $x = 5$

Exercise 2

1 **Solve by multiplying one equation and eliminating *x*.**

a) $2x + 3y = 13$
$x + 2y = 7$

$x = $ ....

$y = $ ....

b) $3x + 2y = 20$
$-x + y = 5$

$x = $ ....

$y = $ ....

2 **Solve by multiplying one equation and eliminating *y*.**

a) $3x + 2y = 19$
$2x - y = 8$

$x = $ ....

$y = $ ....

b) $2x + 3y = 10$
$5x + y = 38$

$x = $ ....

$y = $ ....

3 Solve:

a) $3x + 2y = 26$
$x + 4y = 22$

b) $x - 3y = 15$
$4x + y = 8$

$x = $ ....

$y = $ ....

$x = $ ....

$y = $ ....

4 James buys 5 cakes and 2 bottles of water. He pays $13.

Mila buys 1 cake and 4 bottles of water. She pays $8.

By forming and solving a pair of simultaneous equations, calculate the cost of:

- a cake
- a bottle of water.

Let *c* represent the cost of a cake and *b* represent the cost of a bottle of water.

A cake costs $........... and a bottle of water costs $............

5 Do solutions to simultaneous equations have to be whole numbers?

Can you find two pairs of equations which have solution $x = \frac{1}{2}, y = \frac{1}{3}$?

25.3 Solving simultaneous equations using substitution

Summary of key points

Sometimes the easiest way to solve simultaneous equations is to substitute one equation into the other.

Example:

Solve: $y = x - 1$ ①

$3x + 4y = 24$ ②

Substitute equation ① into equation ② to replace y.

$3x + 4(x - 1) = 24$

$3x + 4x - 4 = 24$

$7x = 28$

$x = 4$

Substitute back into equation ① to find y.

$y = 4 - 1$

$y = 3$

Exercise 3

1 Use substitution to solve these pairs of simultaneous equations.

a) $x = 2$
$x + 3y = 10$

b) $y = x$
$5x + 2y = 56$

$x = $ ....

$y = $ ....

$x = $ ....

$y = $ ....

c) $y = 3$
$2x + y = 10$

d) $y = 2x$
$x + 4y = 36$

$x = $ ....

$y = $ ....

$x = $ ....

$y = $ ....

e) $y = x - 1$
    $5x + y = 23$

f) $y = x + 2$
    $6x + 5y = 21$

$x =$ ....

$y =$ ....

$x =$ ....

$y =$ ....

2 Mateo is trying to solve these simultaneous equations.

$y = x - 4$

$x + 3y = 20$

Here is his working:

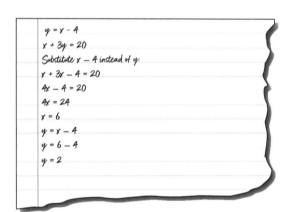

$y = x - 4$
$x + 3y = 20$
Substitute $x - 4$ instead of y:
$x + 3x - 4 = 20$
$4x - 4 = 20$
$4x = 24$
$x = 6$
$y = x - 4$
$y = 6 - 4$
$y = 2$

Is he correct?

....

If he has made a mistake, explain what he has done wrong.

....

....

3 Solve these pairs of simultaneous equations by substitution.

a) $y = 2x + 1$
$4x + 3y = 8$

b) $y = 2x - 5$
$5x - 3y = 13$

$x =$ ....

$y =$ ....

$x =$ ....

$y =$ ....

c) $x + y = 6$
$3x + 2y = 24$

d) $x - y = 5$
$2x + 5y = 24$

$x =$ ....

$y =$ ....

$x =$ ....

$y =$ ....

4 Camille and Dylan have 22 books between them. Camille has 8 more books than Dylan.

a) Write two simultaneous equations to represent how many books they have. Use x for the number of books Camille has and y for the number of books that Dylan has.

....

....

b) Solve your simultaneous equations to find out how many books they have each.

Camille ....

Dylan ....

5 **Freya is thinking of two numbers, a and b.**

When she adds the two numbers she gets 8.

When she doubles the first number and adds four lots of the second number, she gets 22.

a) Write two simultaneous equations in a and b to represent Freya's calculations.

....

....

b) Solve your equations to find the values of a and b.

$a =$ ....

$b =$ ....

6 Amy and Emma have a combined age of 16.

In four years' time, Emma will be twice as old as Amy.

Form and solve a pair of simultaneous equations to work out how old Amy and Emma are.

Amy ....

Emma ....

Make up another problem like this to test out on a partner using Amy and Emma's ages.

25.4 Solving simultaneous equations using graphs

Summary of key points

For the equations

$x + 2y = 7$

$2x - y = 4$

drawing both lines on a coordinate grid shows that the lines cross at (3, 2), so the solution to these simultaneous equations is

$x = 3, y = 2$.

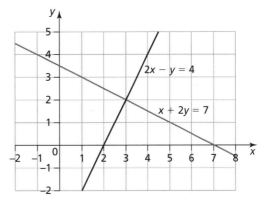

1 Four lines are drawn on the grid.

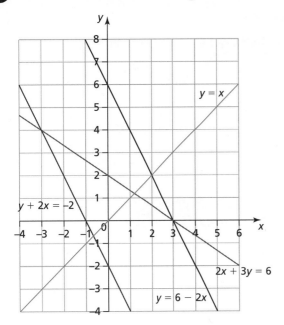

a) Find the solutions to the simultaneous equations

$y = x$ and $y = 6 - 2x$. $x = $ .... , $y = $ ....

b) Find the solutions to the simultaneous equations

$2x + 3y = 6$ and $y = 6 - 2x$. $x = $ .... , $y = $ ....

c) Find the solutions to the simultaneous equations

$2x + 3y = 6$ and $y + 2x = -2$. $x = $ .... , $y = $ ....

d) Find the approximate solutions to the simultaneous equations

$y = x$ and $y + 2x = -2$. $x = $ .... , $y = $ ....

e) Which two lines shown on the grid do not intersect?

.... and ....

2 The graphs of $x = 4$ and $y = 2x - 2$ are drawn on the grid.

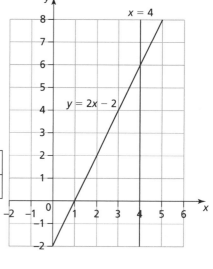

a) Write down the coordinates of the point of intersection of the lines $x = 4$ and $y = 2x - 2$.

(………. , ……….)

b) Complete the table of values for $y = x + 1$ and $y = 5 - 3x$.

$y = x + 1$

x	−1	0	1	2
y				

$y = 5 - 3x$

x	−1	0	1	2
y				

c) Draw the lines $y = x + 1$ and $y = 5 - 3x$ on the grid.

d) Find the solutions to the simultaneous equations

$y = x + 1$ and $y = 5 - 3x$.

$x = $ …………. , $y = $ ………….

e) Find the solutions to the simultaneous equations

$y = x + 1$ and $y = 2x - 2$.

$x = $ …………. , $y = $ ………….

3 a) Complete the tables of values for $x + y = 5$ and $y = 8 - 3x$.

$x + y = 5$

x	0	1	2	3
y				

$y = 8 - 3x$

x	0	1	2	3
y				

b) Plot the graphs of both lines on the grid.

c) Find the solution to the simultaneous equations

$x + y = 5$ and $y = 8 - 3x$.

$x = $ …………. , $y = $ ………….

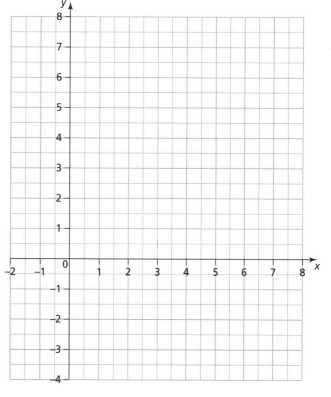

4 **The diagram shows the graph**
of 2x + 3y = 13.

a) Draw the line $2y = 3x - 2$ on the grid.

b) Use your graphs to write down the solutions to the simultaneous equations $2x + 3y + 13$ and $2y = 3x - 2$.

x = ... , y = ...

c) Nina says that the simultaneous equations $2y = 3x - 2$ and $2y - 3x = 10$ have no solution.

Is Nina correct? Yes ☐ No ☐

Explain your answer.

....

....

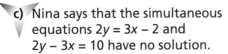

Think about

5 **A pair of simultaneous equations has the solution $x = 3$, $y = -1$.**

One of the equations can be represented by the line $x + 4y = -1$.

Find two possible lines that could represent the other equation.

Thinking statistically

You will practice how to:

- Interpret data, identifying patterns, trends and relationships, within and between data sets, to answer statistical questions. Make informal inferences and generalisations, identifying wrong or misleading information.
- Record, organise and represent categorical, discrete and continuous data. Choose and explain which representation to use in a given situation:
  - o Venn and Carroll diagrams
  - o tally charts, frequency tables and two-way tables
  - o dual and compound bar charts
  - o pie charts
  - o line graphs, time series graphs and frequency polygons
  - o scatter graphs
  - o stem-and-leaf and back-to-back stem-and-leaf diagrams
  - o infographics.

26.1 Misleading graphs

Exercise 1

1 **Which of the following makes this graph misleading:**

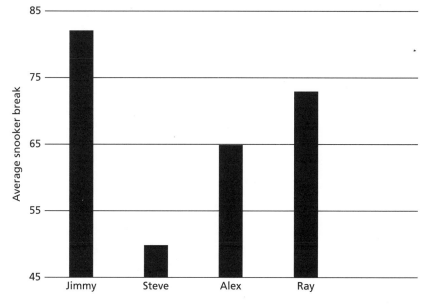

	True	False
y-axis does not start at zero		
No units on the *y*-axis		
y-axis numbers do not increase evenly		

2 **The pie charts show the number of sales of hand sanitiser by two companies.**

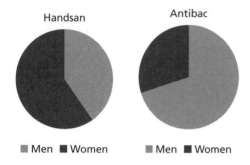

Handsan   Antibac

■ Men ■ Women   ■ Men ■ Women

Decide if the statements are True, False or Unsure.

	True	False	Unsure
More women buy Handsan than men			
More women buy Handsan than Antibac			
More men buy Handsan than Antibac			

3 **Write down what is incorrect about this chart.**

SuperMart–Revenue by sector

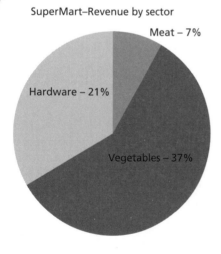

Meat – 7%

Hardware – 21%

Vegetables – 37%

....

....

....

4 What is misleading about the graph and the statement 'Sharp decline in city rat population'?

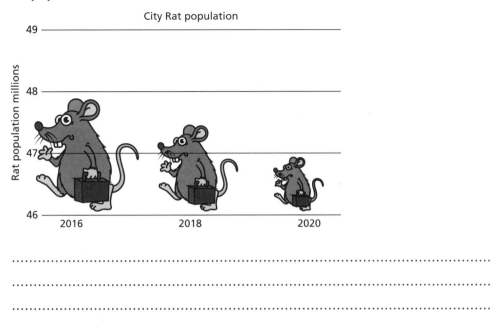

....

....

....

5 **Think about**

Compare the area of each rat image with the actual decrease in population.